High Resolution Optical Satellite Imagery

High Resolution Optical Satellite Imagery

Ian Dowman
Karsten Jacobsen
Gottfried Konecny
Rainer Sandau

Whittles Publishing

Published by
Whittles Publishing,
Dunbeath,
Caithness KW6 6EG,
Scotland, UK

www.whittlespublishing.com

Distributed in North America by
CRC Press LLC,
Taylor and Francis Group,
6000 Broken Sound Parkway NW, Suite 300,
Boca Raton, FL 33487, USA

ISBN 978-184995-046-6

Printed by Almarose, Slovenia

Contents

Preface

Photogrammetry developed slowly during the first 70 years of the twentieth century, but since 1972, when ERTS – later to become Landsat – was launched, the changes have become rapid. Landsat introduced digital images to a wide audience, and when SPOT-1 – an Earth observation satellite – was launched in 1986, photogrammetry using images from space became a major interest of mapmakers and scientists. With Landsat, remote sensing became a recognised subject, initially concerned with the interpretation and classification of images, but then converging with photogrammetry, so that today the two subjects are fully integrated. This is particularly so with high resolution optical images, which compete with aerial imagery for smaller-scale mapping; high resolution data is also used for interpretation, studying the environment and intelligence gathering. This book is concerned with the photogrammetric use of high resolution images, although aspects of image processing are also discussed. The book aims to bring together information on a range of sensors, including their characteristics and the applications to which they are put.

The definition of "high resolution" is not universally agreed upon and for this reason we start with Advanced Spaceborne Thermal Emission and Reflection Radiometer (ASTER) data, with 15 m ground sample distance (GSD). Stereoscopic images are acquired and the data is used photogrammetrically, particularly for digital elevation model (DEM) generation. SPOT images with 10 m GSD are among the most widely used, and established the interest in the use of stereoscopic images from space. The current commercial sensors with sub 1 m GSD are dealt with in the most detail because of their current importance. Photographic images are not forgotten, however, although their use today is minimal, and is mainly for monitoring purposes. We aim to present comprehensive information on all of these sensors and put them into context in the current world of geospatial information.

The objective of the book is to explain the way in which high resolution imagery is obtained – along with the issues which surround its collection and use – to students, people working in map production and scientists working in fields other than mapping. A step-by-step approach is employed, beginning with a discussion of the fundamentals of mapping from imagery in Chapter 1.

Chapter 2 presents a historical overview of the development of optical sensors in space and covers CORONA and other film cameras. These played an important role in the establishment of the use of image data from space, but the problems of film recovery and the limitations of film in processing images led to their demise once SPOT demonstrated what could be done with push-broom stereo imagery.

Chapter 3 deals with the generic aspects of sensor hardware and the systems needed to record and transmit data from space, with an introduction to instrumentation to provide data for direct georeferencing of images from space.

Chapters 4 and 5 give details of specific sensors, particularly those in widespread use. Chapter 4 covers sensors with a GSD of greater than 1 m and includes SPOT 1–4 and ASTER; these sensors are mostly designed to collect stereoscopic data using two or more telescopes, pointing at fixed or variable angles. On the other hand, the sensors with a GSD of greater than 1 m, discussed in Chapter 5, are mainly agile sensors which can be pointed in any direction, up to given limits. These two chapters discuss the technical specifications and performance of the sensors, based on published studies.

Chapter 6 is devoted to with calibration, sensor models and orientation. This is a key chapter for users of the data and covers both rigorous models and rational polynomial coefficients, as well as variations of these methods. This chapter ends with a comparison of orientation methods.

Chapter 7 deals with product generation; the discussion is fairly generic and includes the principles of image matching and the particular problems of DEM generation from satellite-borne sensors. Issues of resolution and interpretation are discussed, again in the context of satellite data.

Finally, Chapter 8 presents a review of future developments, discussing future missions and the issues which face further development.

The authors gratefully acknowledge support from organisations that have provided information and material used in preparing this book, in particular SSTL of Guildford (UK), GeoEye and DigitalGlobe. We also thank Keith Whittles for his encouragement to start and finish the book and Manos Baltsavias and Armin Gruen for their helpful comments.

We hope that this book will help to expand the use of high resolution satellite data by enabling students and professionals to understand the technology involved and to appreciate and wonder at the magnificent images and detailed information which can be obtained from hundreds of kilometres above the surface of the Earth.

Ian Dowman
University College London
Karsten Jacobsen
IPI, Leibniz University, Hannover
Gottfried Konecny
IPI, Leibniz University, Hannover
Rainer Sandau
German Aerospace Center

The Authors

Ian Dowman has spent most of his career working at University College London, initially carrying out research with aerial and close range photography, but moving towards an interest in the application of digital aerial and satellite data to mapping. He has been involved in many projects involving the use of satellite data for mapping. He is now retired from UCL but continues to teach and have an active interest in international affairs. Ian was president of ISPRS from 2004–2008 and is now First Vice President.

Dr. Karsten Jacobsen is retired as Academic Director of Institute of Photogrammetry and GeoInformation, Leibniz University Hannover; his main field of research is numerical photogrammetry, handling of optical aerial and space images, and determination of digital elevation models. He is a leading authority on modeling and processing satellite images.

Professor Gottfried Konecny is very well known as a research scientist and educator principally through his career at the University of New Brunswick in Canada and Leibniz University, Hannover. He has four Honorary Doctorates and has led and been involved in many projects involving mapping from space, particularly the Metric Camera Experiment on Space Shuttle for the European Space Agency and the MOMS sensor on Space Shuttle and the MIR Space Station. He has been a consultant in projects throughout the world and was President of International Society for Photogrammetry and Remote Sensing 1984-88 and Chairman of European Association of Remote Sensing Laboratories 1993-97.

Dr.-Ing. habil. Rainer Sandau is retired from the German Aerospace Center (DLR) and but acts currently in numerous positions including adjunct Professor of Baylor University, Texas, USA, and Technical Director Satellites and Space Applications of the International Academy of Astronautics (IAA). He has over 30 years of experience in airborne and spaceborne activities and has led and been involved in instrumentations of four space missions to Venus, Mars and Earth, and also numerous concepts for instruments and small satellites for or with different countries and space agencies. He authored or co-authored over 250 publications, holds over 25 patents, and is co-author or editor/co-editor of 21 books or conference proceedings.

List of Abbreviations

ADEOS	Advanced Earth Observing Satellite
ALOS	Advanced Land Observing Satellite
ASTER	Advanced Spaceborne Thermal Emission and Reflection Radiometer
AVHRR	Advanced Very High Resolution Radiometer
AVNIR	Advanced Visible and Near-Infrared Radiometer
AWiFS	Advanced Wide Field Sensor (ISRO)
B:H	base-to-height ratio
BER	bit error rate
BIRD	bi-spectral infrared detection
BLMIT	Beijing Landview Mapping Information Technology Co
BNSC	British National Space Centre
bps	bits per second
BPSK	binary phase shift keying
BRDF	bi-directional reflectance distribution function
CCD	charge-coupled device
CE90	circular error at the 90% probability level
CEOS	Committee on Earth Observation Satellites
CNES	Centre National d'Etudes Spatiales (France)
CNSA	China National Space Administration
DCT	discrete cosine transformation
DEM	digital elevation model
DLR	German Aerospace Centre
DLT	direct linear transformation
DMC	Disaster Monitoring Constellation (a constellation of the satellites produced by SSTL)
DOQQ	digital orthophoto quarter quads
DORIS	Doppler satellite-tracking/positioning system (France)

DP	dynamic programming (used for pixel-based matching)
DR	dynamic rate (sensor rate measured in bps)
DRA	dynamic range adjustment
DS	dark signal
DSM	digital surface model (height model of visible surface)
DSNU	dark signal non-uniformity
DTM	digital terrain model (height of bare ground)
DV	data volume
EADS	European Aeronautic Defence and Space Company
ECEF	Earth-centred Earth-fixed (coordinate system)
ERTS-1	Earth Resources Technology Satellite-1, renamed Landsat-1
ESA	European Space Agency
ETM	extended thematic mapper
f	focal length of a lens or optical system, also known as "principal distance"
FOV	field of view
FPM	focal plane module
Gb	gigabit
GB	gigabyte
GCP	ground control point
GDEM	(ASTER) global digital elevation model
GEOSS	Global Earth Observation System of Systems
GeoTiff	a format for presenting images with geographical coordinates
GIS	graphical information system
GISTDA	Geo-Informatics and Space Technology Development Agency (Thailand)
GMES	Global Monitoring for Environment and Security
GNSS	global navigation satellite system
GPS	global positioning system (the US version of GNSS)
GSD	ground sample distance
GTOPO30	worldwide elevation model compiled by USGS with 30-arcsec spacing
HRG	high resolution geometric (sensor on SPOT-5)
HRS	high resolution stereo (stereo sensor on SPOT-5)
HRV	high resolution visible (sensor on SPOT 1–4)
i	incidence angle; can also stand for inclination (angle of the plane of an orbit with the equator)
ICP	independent check point
IFOV	instantaneous field of view
IfSAR/InSAR	interferometric synthetic aperture radar
IFT	inverse Fourier transform

IHS	intensity, hue, saturation
INPE	National Institute for Space Research (Brazil)
INS	inertial navigation system
IPG	Institute for Photogrammetry and GeoInformation (Hannover University)
IR	infrared
ISF	intensity spread function
ISPRS	International Society of Photogrammetry and Remote Sensing
ISRO	Indian Space Research Organisation
ITRF	international terrestrial reference frame
ITRS	international terrestrial reference system
JAXA	Japan Aerospace Exploration Agency
JPEG	Joint Photographic Experts Group (a standard for compression of data)
KH	US Keyhole reconnaissance satellites
KLM	a file format used to display geographic data in an Earth browser such as Google Earth, Google Maps, and Google Maps for mobile
LE90	linear error at the 90% probability level, used for height
LEO	low earth orbit
LFC	large format camera (analogue film camera with 23 cm × 46 cm film format)
lidar	light detection and ranging
LISS	linear self-scanning sensors (ISRO)
KARI	Korea Aerospace Research Institute (South Korea)
Mb	megabit
MB	megabyte
MC	metric camera (analogue film camera with 23 cm × 23 cm film format)
METI	Ministry of Economy, Trade and Industry (Japan)
MOMS	Modular Optoelectronic Multispectral Stereo Scanner
MTF	modulation transfer function
MI	mutual information
MS	multispectral
ν	nadir angle
NARA	National Archives and Records Administration
NASA	National Aeronautics and Space Administration
NASDA	National Space Development Agency of Japan
NGA	National GeoSpatial Intelligence Agency
NIR	near infrared

NOAA	National Oceanic and Atmospheric Administration (US Federal Agency)
NRO	National Reconnaissance Office
NSPO	National Space Programme Office (Taiwan)
NSSDC	National Space Science Data Center
O	perspective centre or projection centre – the point through which all rays pass for a central projection or perspective projection
PALSAR	Phased Array type L band Synthetic Aperture Radar
PAN	panchromatic
PCA	principal component analysis
Period	time taken for a satellite to complete one orbit
PPM	piecewise polynomial model
PRISM	Panchromatic Remote-sensing Instrument for Stereo Mapping
PRNU	photo response non-uniformities
PSF	point spread function
QPSK	quadrature phase shift keying
RAL	Rutherford Appleton Laboratory
RGB	red, green and blue
rms	root mean square
rmse	root mean square error
RPC	rational polynomial coefficients
σ	standard deviation
SAC	Space Application Centre (India)
SAR	synthetic aperture radar
SGM	semi-global matching
SNR	signal-to-noise ratio
SPOT	*satellite pour observation de terre* (Earth observation satellite)
SRTM	Shuttle Radar Topography Mission (US mission on the space shuttle to generate a global DSM by InSAR)
SST	stereo strip triangulation
SSTL	Surrey Satellite Technology Ltd (UK)
SWIR	short-wave infrared (mid infrared)
TDI	transfer/time delay and integration
TIN	triangulated irregular network
TIR	thermal infrared
TMA	three-mirror anastigmatic
TÜBITAK	Scientific and Technological Research Council of Turkey
USGS	United States Geological Survey
UTM	Universal Transverse Mercator

VLL	vertical line locus
VNIR	visible and near infrared
WGS84	World Geodetic System 1984
WiFS	Wide Field Sensor (ISRO)
WMS	OpenGIS® Web Map Service Interface Standard (provides a simple HTTP interface for requesting geo-registered map images from one or more distributed geospatial databases)

Satellite Imaging Technology 1

Ian Dowman

1.1 Introduction

Imagery acquired from sensors mounted on a satellite has been available to the general public since 1972, when ERTS-1, later to become Landsat-1, was launched. Since then the resolution has steadily increased. The first big jump was with SPOT-1 HRV in 1986, which was followed by IKONOS in 1999. These developments have led to the widespread use of satellite data for mapping and the generation of three-dimensional (3D) data. A great deal of information on the methods of using such data and on the results achieved can be found in journals, but this book aims to give a complete overview of high resolution optical sensors operating from space and the way in which the data can be processed and used.

We will present some basic principles and show how they have been applied to current sensors. The push-broom imaging system has been used in most sensors launched since SPOT-1 and this has therefore been taken as the common factor. So, for the purpose of this book, "high resolution" is taken to include sensors with a pixel size of less than 16 m. This limit is adopted in order to include SPOT and ASTER (Advanced Spaceborne Thermal Emission and Reflection Radiometer), which are important sensors for mapping and DEM generation. Classification of sensors is discussed in Section 1.2. These sensors use push-broom technology, can derive images from off-nadir viewing and can be calibrated and used for mapping. Landsat and similar sensors are not covered, as these are not normally rigorously oriented and are mainly used for deriving environmental data. This first chapter will give an introduction to SAR and InSAR but will not go into detail. Lidar is not discussed, as its use from space is very limited.

1.2 Definitions

A number of terms are used throughout the book; the important ones are defined and discussed here. Abbreviations and commonly used terms are given in the List of Abbreviations.

1.2.1 Resolution and pixel size

Resolution is a measure of what can be identified on an image. There are several measures of this, which are discussed in detail by Forshaw *et al.* (1983), and the subject is covered in Section 3.3.1. A simple measure which has been used for many years for determining resolution of a film camera is "line pairs per millimetre", or lp mm^{-1}. A target is constructed and photographed, and the greatest number of lines which can be identified on the image defines the resolution. This can be compared to the resolution of the human eye, which is 10 lp mm^{-1}, whereas the resolution of a typical aerial photograph is 40 lp mm^{-1}; the latter is equivalent to 25 μm lp^{-1}, which equates to a pixel size of 9 μm on the image. The relationship is given by Light (1990)

$$\mu m\ 1p^{-1} = (\text{pixel size at image scale}) / (2\sqrt{2}) \quad (1.1)$$

Jacobsen (2010) gives a more up to date relationship between analogue and digital imaging. The pixel size is not the same as resolution defined by μm lp^{-1}, although resolution is frequently quoted as the pixel size. The pixel size is related to the instantaneous field of view (IFOV) for a digital sensor. A pixel is the cell in a raster image (Figure 1.1).

Ground sample distance (GSD) will be used in this book to define the resolution of a sensor. Topan *et al.* (2005) define GSD as: "... the distance of the centres of neighbouring pixels projected to the ground. The pixel size on the ground is the physical size of the projected pixels." They also note that "pixels may be over-sampled (the projected pixels are overlapping) or under-sampled (there is a gap between neighbouring pixels)." The pixel size in the along-track direction is defined by the sampling interval. As an example they quote SPOT-5, which generates images with 2.5 m *GSD* in the supermode images, but the neighbouring pixels overlap by 50%, so the pixel size on the

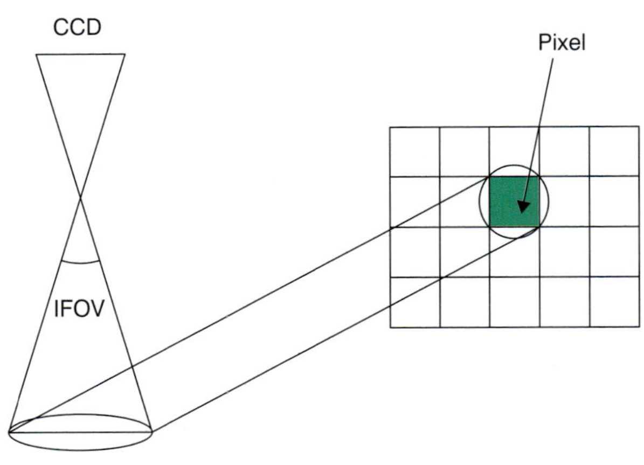

Figure 1.1 Instantaneous field of view and pixel size.

ground is 5 m. In theory, the information content of the SPOT-5 supermode corresponds to an effective *GSD* of 3 m. *GSD* can be defined as

$$GSD = ISF \, . \, (H/f) \qquad (1.2)$$

where *ISF* is the point spread function (PSF), which is dependent on the optical system and the atmosphere. Light (2004) replaces *ISF* with the pixel size in the sensor; this topic is discussed further in relation to current sensors by Thomson (2010). The *GSD* determined by practical tests will normally be greater than the *GSD* quoted as a theoretical measure. Where *GSD* is quoted in this book without qualification, it will refer to the nominal value determined directly from the IFOV. Topan *et al.* (2005) also discuss this in some detail.

1.2.2 Classification of sensors

As explained above, for the purpose of this book, the term "high resolution" refers to sensors with a pixel size of less than 16 m, but this is by no means a universal classification. Schreier and Dech (2005), for example, class anything from 1 to 30 m as "high resolution", and satellites with a pixel size of less than 1 m as "very high resolution". The following classification based on pixel size has also been widely used

Low resolution	≥ 30 m and < 300 m
Medium resolution	≥ 5 m and < 30 m
High resolution	≥ 1.0 m and < 5 m
Very high resolution	< 1.0 m

However, this book will take high resolution as defined earlier, but will include other satellites in the range ≥ 0.5 m but < 16 m, for reasons explained above.

The term "small satellites" is also used fairly loosely; Table 1.1 gives a classification of platforms [Xue *et al.*, 2008]. Optical satellites are often grouped by their weight as shown in Table 1.1, but this grouping is not universally accepted. With advanced components, satellites may be lighter today than satellites of a few years ago with the same

Table 1.1 Classification of small satellites

Small satellite (kg)				Large satellite (kg)
PICO	Nano	Micro	Mini	
> 0.1–1	> 1–10	> 10–100	> 100–1000	> 1000–10 000

specification. Several components are available on the market, leading to cost reductions and enabling several countries to build or even buy a satellite and operate it themselves.

1.2.3 Accuracy and precision

It is always important to distinguish between accuracy and precision. Within mapping circles, accuracy is often quoted in terms of the root mean square error (rmse) which compares computed coordinates with reference coordinates, usually ground control points (GCPs) or ground check points and is determined by the equation

$$rmse = \sqrt{\sum v^2 / n} \tag{1.3}$$

where v is the difference between the computed coordinate and the reference coordinate, usually referred to as a "residual", and n is the number of points.

The literature contains a confusing mixture of terms, particularly "standard deviation" (σ), which has a probability level of 68%, and "circular error", where a 90% probability (CE90) is used. In the normal, or Gaussian, distribution there is the factor 1.65 between 90% and 68% probability levels (Figure 1.2). The circular error is given by $(\sigma_x^2 + \sigma_y^2)^{½}$. Under the condition of equal σ_x and σ_y, the circular error CE90 is identical

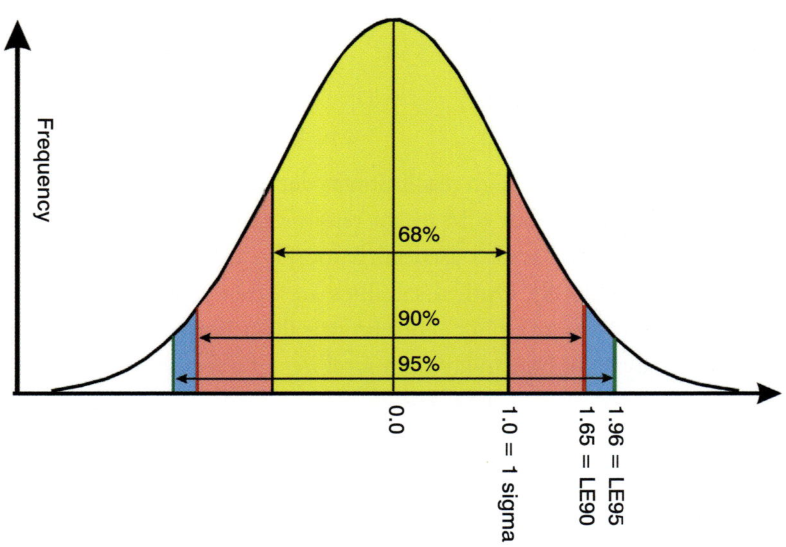

Figure 1.2 The normal (or Gaussian) distribution, which gives the frequency of error distribution in relation to the standard deviation of the height (σ_z) or one sigma, to the linear error LE90 and LE95.

to σ_x . 2.146, respecting the reduced probability of large values for X and at the same time for Y. For the vertical component the linear error with 90% probability (LE90) corresponds to the vertical standard deviation (σ_z) times 1.65 *rmse*.

1.2.4 Geolocation, georeferencing and geocoding

Images must be related to a reference system in order to be used for mapping. Most countries have their own national reference system based on a map projection, and there are also international reference systems, such as Universal Transverse Mercator (UTM). More details of these can be found in Iliffe and Lott (2008). Images may be related to a reference system with varying degrees of accuracy and a number of terms are used to refer to this. "Geolocation" is a term widely used in the GIS community, especially when referring to web-delivered data. There is no accepted definition; neither is there any defined accuracy linked to this. The following terminology will be used in this book:

- **Geolocation** will indicate an approximate location of an image, without any correction being applied.
- **Georeferencing and geopositioning** imply greater accuracy than geolocation. We will take a georeferenced image to be one that has been corrected for all distortion except relief, and from which approximate coordinates of a point, in a projection or latitude and longitude, can be found.
- **A geocoded image** is one in which relief distortion has also been corrected and is referred to a defined coordinate system with good accuracy. The term is synonymous with "orthoimage".

Rigorous orientation to a reference system using photogrammetric methods is referred to as "exterior orientation", and careful attention has to be given to reference systems when carrying this out. The object coordinate system is normally a 3D right-handed, Earth-centred fixed-coordinate system relative to the Earth and rotating in space. This can be an International Terrestrial Reference System (ITRS), WGS84 or any other Earth-centred system. ITRS is maintained to take account of tectonic movements of the Earth and the ITRF website provides transformation parameters to other systems [ITRF, 2011]. National reference systems are defined to meet the requirements of a particular country in order to allow the most accurate representation of the area (see Section 1.3.7).

1.2.5 Digital elevation models

A digital elevation model (DEM), also referred to as a "digital height model", is a common product derived from stereoscopic images and relates to a "cloud" of points with X, Y, Z coordinates; these points are usually on the ground or on features of vegetation on the surface of the ground. 3D coordinates are computed from the stereoscopic images of points on the ground, and the resulting set of randomly distributed points are used to define the surface of the ground, either by a triangulated irregular network (TIN) or by a grid. A digital surface model (DSM) includes the features on the surface of the Earth and a digital terrain model (DTM) defines the terrain surface. More on DEMs can be found in Chapter 7.

1.2.6 Orthoimages

All images of the Earth are subject to distortion due to the relief of the ground (see Section 1.3). An orthoimage has been corrected for this distortion and displays the image in a defined coordinate system, as in a map. Orthoimages are discussed in Section 7.5.

1.2.7 Image products

The organisations collecting and distributing image data use different names for their products. "Spot Image" is a typical example as shown in Table 1.2.

Table 1.2 SPOT scene products (adapted from http://www.spotimage.com/automne_modules_files/standard/public/p234_0916b39e58c710b3bb5788fdcc025f80niveau_anglais_2008.pdf)

Level	Description
Level 1A	Level 1A imagery is corrected by normalizing CCD response to compensate for radiometric variations due to detector sensitivity. Only geometric corrections according to the inner orientation are performed. Level 1A pre-processing thus leaves data in almost raw form.
Level 1B	Level 1B applies the same radiometric corrections as level 1A. Geometric corrections compensate for systematic effects, including panoramic distortion, the Earth's rotation and curvature, and variations in the satellite's orbital altitude.
Level 2A	Level 2A scenes are rectified to match a standard map projection (UTM WGS 84), without using GCPs. Level 2A is the entry-level map product.
Level 2B	Level 2B imagery is georeferenced, i.e. scenes are framed in a given map projection and tied to GCPs (obtained from a map or topographic surveys) for even better location accuracy.
Level 3	Level 3 imagery is georeferenced like level 2B. Level 3 products, also called "orthoimages", are pre-processed using a DEM to correct residual parallax errors due to relief.

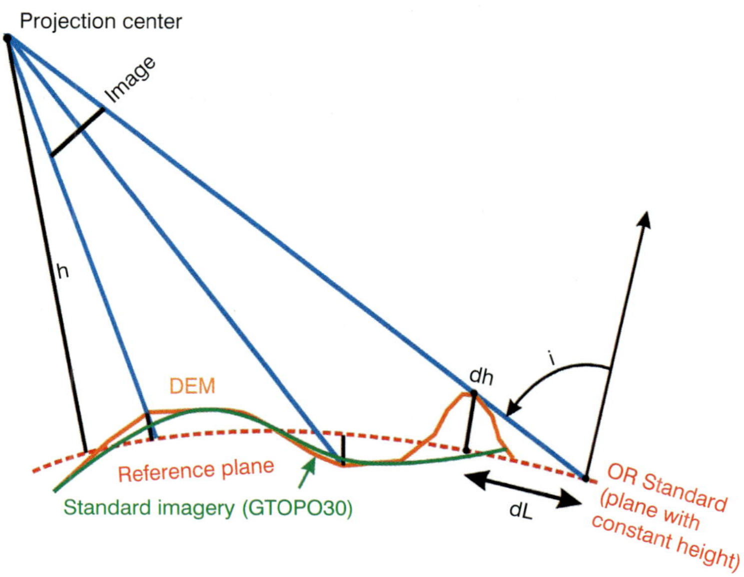

Figure 1.3 Errors from using DEMs of different accuracy.

The high resolution data companies tend to use a more application-oriented approach. For example, GeoEye uses the product names of Geo, GeoProfessional and GeoProfessional Precision [GeoEye, 2011]. All of these are georeferenced with specified precision and are roughly equivalent to level 2A, level 2B and level 3 in the SPOT system. It is important to understand the product specification before ordering or using satellite images. These products are discussed further in the sections on particular sensors.

Some standard products are corrected with DEMs of an accuracy that is not good enough to be used as height reference for an orthoimage with a ground pixel size of less than 1 m; for example, QuickBird images are corrected with GTOPO30 (see Section 7.4.1). The location error caused by an insufficient reference height model for an orthoimage is $dL = dh \,.\, \tan i$, where i is the incidence angle and dh is the error of the height model as shown in Figure 1.3.

1.3 Outline of principles

1.3.1 Satellite orbits

It is necessary have an understanding of satellite orbits to appreciate the characteristics of high resolution satellite data. Further details can be found in Montenbruck and Gill (2000). A satellite orbits the Earth under the force of gravity, according to Kepler's law. The general shape of an orbit is elliptical, but most satellites carrying an imaging payload have a near-circular orbit. The plane of the orbit makes an angle with the equator,

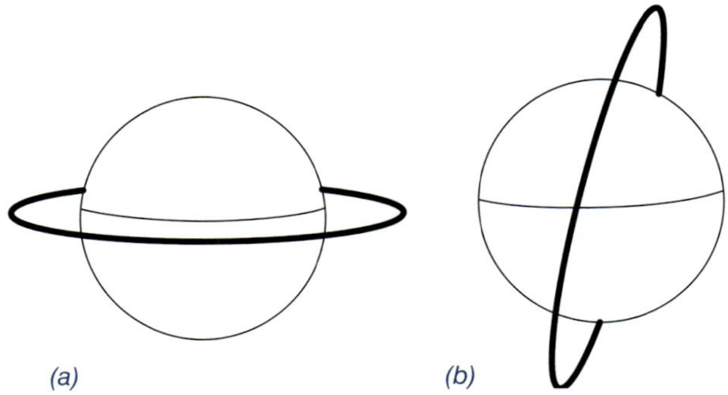

Figure 1.4 Earth orbits: (a) equatorial orbit, inclination 0°; and (b) near-polar orbit, inclination 98°.

called the "inclination". A zero inclination indicates an equatorial orbit; an orbit with a 90° inclination is a polar orbit. Imaging satellites mostly have a sun-synchronous, near-polar orbit with an inclination of about 98° as shown in Figure 1.4. Orbiting at an altitude of between 700 and 800 km, the satellite in a near-polar orbit sees virtually every part of the Earth as the Earth rotates underneath it. The satellite makes a complete orbit in approximately 90 minutes – this is called the "period". Near-polar orbits allow a satellite to pass over a section of the Earth at the same time of day.

As an example, the characteristics of the IKONOS orbit (which is typical of Earth-observing satellites) are

Inclination	98.1°, sun synchronous
Period	97 minutes
Equator crossing time	nominally 10:30 a.m. solar time
Altitude	681 km

GeoEye-1 has an orbital separation on the equator of 2694 km and a theoretical repeat cycle of 177 days. With this orbit, every point on the surface of the Earth can be imaged every three days by using a nadir angle of 30°, and every day by using an angle of 60°. The coverage of IKONOS in 24 hours is given in Figure 1.5.

The DigitalGlobe constellation of QuickBird, WorldView-1 and WorldView-2, with up to 30° off-nadir view, gives access to locations at 28° latitude on 28 occasions within a 30-day period. On some days there can be three passes and the maximum gap between passes is three days.

1.3.2 Geometry of a single image

In order to produce 3D information from pairs of images, it is necessary a) to know the geometry of the image-forming system; b) to consider the movement of the platform on

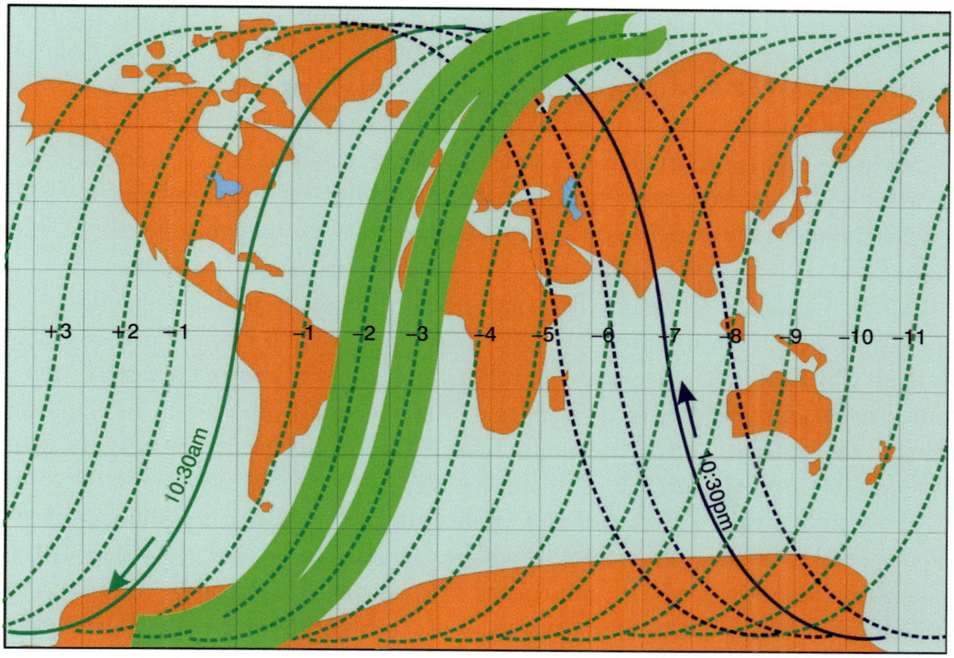

Figure 1.5 IKONOS orbits in 24 hours: blue lines indicate the ascending orbit (in darkness), and light-green bands indicate the approximate range of imaging with a nadir angle of up to 50°.

which the sensor is mounted; and c) to define the shape of the ground which is covered by the image. The importance of each of these factors will depend on the type of sensor used and the path which the sensor is following relative to the ground. The detailed effect of these factors with respect to the different sensors must be discussed in the context of the sensors themselves. A discussion of effects common to all sensors is given here.

An image may be formed in three ways:

1. As a single perspective exposure, that is with no significant movement of the sensor while the image is formed, as in a frame camera. In this case there is no need to introduce a time parameter into the mathematical model. The model for a whole frame is the central projection and is the normal method used in photogrammetric restitution.
2. As a series of lines. In this case a single line can be considered without a time parameter, but time must be considered in constructing a full frame. The push-broom scanners fall into this category.
3. As a series of points, or groups of points, each recorded at a separate time. This requires the most complex mathematical model. The whisk-broom scanner systems, such as Landsat, fall into this category.

The central projection, also known as a "perspective projection", shown in Figure 1.6 is important because it is the basis of the image formation of a frame in many sensors; it also applies to lines of data in a push-broom system and to individual pixels. When a single frame is formed at an instant of time, a bundle of rays forms the central projection; when a frame is formed from a series of lines, each set of lines is a pencil of rays; even the third case above can be considered as a time-dependent central projection. An understanding of the central projection is also necessary to study the effect of relief on images.

A frame image is formed by light rays from an object being focused onto a focal plane by a lens or via a mirror telescope. The optical system is simplified as a point through which all light rays pass, so that the process of image formation can be represented in diagrammatic form by a series of straight lines from objects A–D, passing through the vertex or perspective centre O onto the focal plane, where images a–d are formed as shown in Figure 1.6.

If the ground were a plane that was parallel to the focal plane, then an image that is a direct scale reduction of the object would be found in the focal plane. The amount of reduction would be given by the ratio of the object distance to the image distance, f/H_a, where f is the principal distance and H_a the sensor altitude above ground. This relationship is used to define the pixel size on the ground. In practice this does not occur and the image is affected by the relief of the ground and the tilt of the sensor; however, an approximate scale factor can be applied.

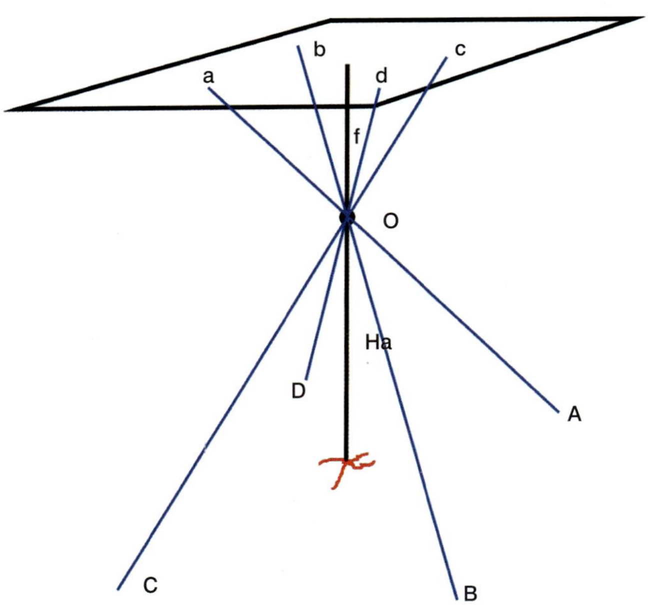

Figure 1.6 The central projection.

Relief has an effect of causing the scale to change throughout the image and also causes images of objects to be displaced, radially from the nadir, from the position at which they should appear if there was no relief. In Figure 1.7 a point A on the surface of the ground should appear on a map of scale f/H at a′ but actually appears on the photograph at a. The magnitude of this displacement is given by the expression

$$aa' = \frac{an.h}{H} \tag{1.4}$$

or in terms of a radial distance from the nadir points (r = na, dr = aa′)

$$\mathrm{d}r = \frac{r.h}{H} \tag{1.5}$$

This can be rewritten as

$$\mathrm{d}r = tan\, v\,.\,h\,.\,\frac{f}{H} \tag{1.6}$$

where v is the nadir angle. This expression indicates that the effect of relief displacement is less for smaller nadir angles. On the edge of a Landsat image, the effect on the ground will be 100 m, or about 3.4 pixels. For normal relief, the effect on Landsat images can be ignored, but for photographic and push-broom systems which look obliquely, it cannot be ignored and 3D geometry must be considered.

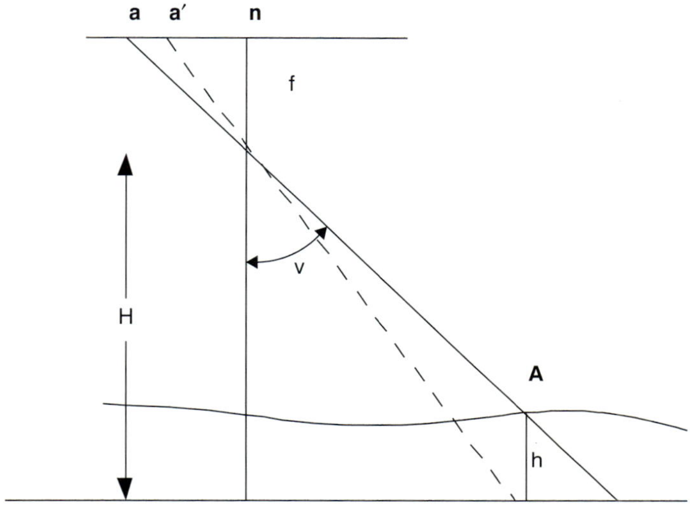

Figure 1.7 Displacement due to relief on a vertical image.

When the interior and exterior orientation is known, the effect of relief can be removed by having a DEM available in order to compute, and hence correct, the displacement caused by relief at every point. An image in which relief displacement has been corrected and pixels are located in a regular object coordinate raster is called an "orthoimage".

In the case of single exposures the platform can be considered to be stationary at the time of exposure, but if the sensor is tilted the pixel size will change over the image. A similar pattern of displacement due to tilt will also be built up as an image is being formed with the push-broom system and there will be cumulative affects due to changes in platform height.

The effects of tilt can be considered in rigorous mathematical terms and the movement of the platform modelled by time-dependent exterior orientation. A full description of photogrammetric image restitution can be found in Mikhail *et al.* (2001) and is discussed in Chapter 6 for satellite images.

1.3.3 Earth effects

When an image is recorded from above the surface of the Earth, the curvature of the surface has to be taken into account. On a map this will have been allowed for by using a map projection [see Iliffe and Lott, 2008]. The effect of Earth curvature on an image will be a reduction of scale towards the edges, so that features appear to be receding into the distance.

Refraction through the atmosphere may also cause distortion but, because the Earth's atmosphere covers only the remote portion of the path of light between the sensor and the surface, this effect is usually small or negligible for satellite images. This is considered further in Section 6.1.

As well as the movement of the satellite, the Earth rotates relative to the satellite. If an image is recorded as a single frame with perspective geometry, then all motion is frozen. If the image is recorded point by point, or line by line, with time intervals, distortion will occur. In push-broom systems the image geometry is affected, but modern agile satellites can compensate for the Earth rotation effect by permanent change of the view direction. SPOT-5 respects the Earth's rotation by continuous rotation of the mirror, known as "yaw correction", which is illustrated in Figure 1.8.

1.3.4 Acquisition of stereoscopic data

If images from two different positions are available, the pair of images can be viewed stereoscopically to give a 3D view of the terrain, and measurements can be made to give 3D object coordinates.

To obtain a stereoscopic image from a pair of images, certain conditions must be satisfied. The sensors must not be too close together or too far apart. If they are too close together, the view will not be different enough to enable a stereoscopic image to

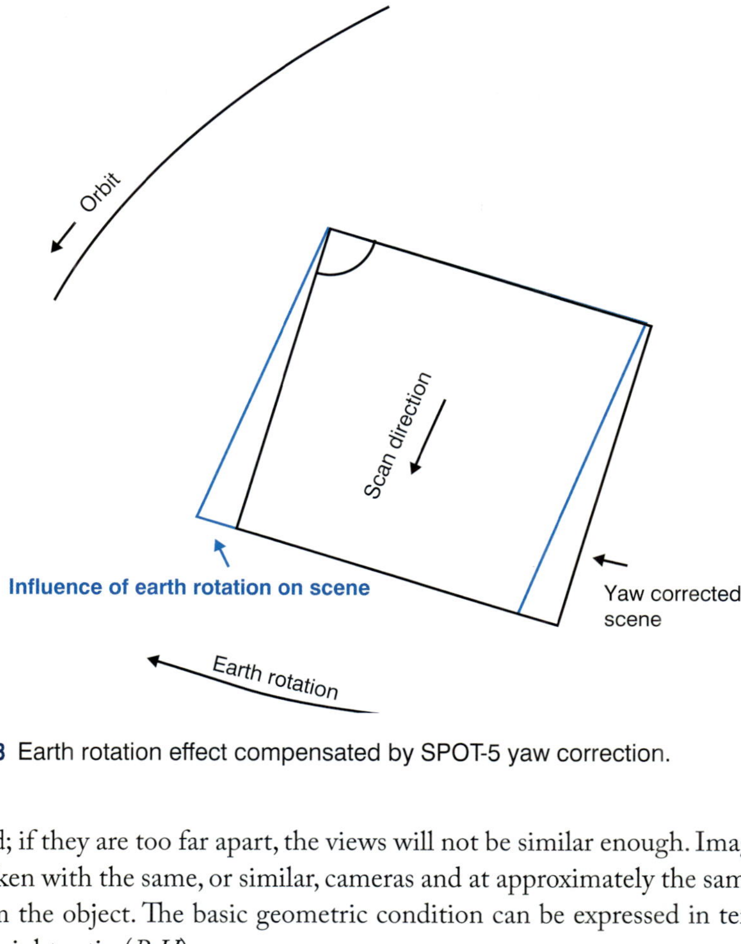

Figure 1.8 Earth rotation effect compensated by SPOT-5 yaw correction.

be formed; if they are too far apart, the views will not be similar enough. Images should also be taken with the same, or similar, cameras and at approximately the same distance away from the object. The basic geometric condition can be expressed in terms of the base-to-height ratio (*B*:*H*).

In Figure 1.9 the distance between the cameras is the base *B*, and the distance between the base and the object is the height *H*. In aerial photography the distance is equivalent to the flying height above ground, *H*. Images taken with a camera from an aircraft for the purpose of constructing a map are carefully controlled so that the axis of the camera is pointing almost vertically downwards and each photograph overlaps its neighbour by 60%. This ensures complete coverage of the ground and a satisfactory base-to-height ratio.

Images from other sensors are generally not controlled in this way. For example, Landsat images, except in high northern and southern latitudes, overlap by only a small amount and cannot be used for stereoscopic viewing. Images from SPOT, however, can be controlled with the view direction to give very good stereoscopic viewing conditions from neighbouring orbits, with nearly 100% overlap.

Images which can be viewed stereoscopically may be obtained in several ways. Fore and aft stereoscopic imagery is the type of acquisition whereby images are formed sequentially

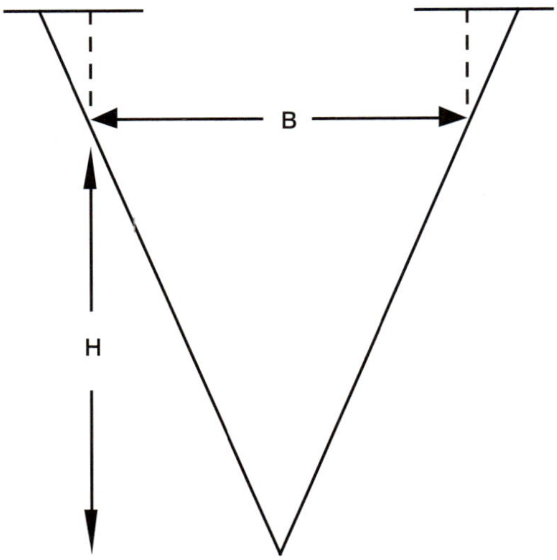

Figure 1.9 Base-to-height ratio.

along the path of the aircraft or satellite. This includes the case of aerial photography overlapping by 60% or more, but can also include a system of linear arrays – one pointing forwards, one vertically and one aft (Figure 1.10a); two sensors can also be used. ASTER and SPOT-5 HRS sensors use fore and aft stereo.

Lateral images can also form stereo pairs. In this method the sensor points downwards or obliquely to one side so that the area of interest is covered on different passes (Figure 1.10b). The disadvantage of this method is that there is a time interval between images during which land cover or illumination conditions may have changed. This is the method adopted by SPOT HRV and HRG as well as other sensors such as IRS-1C and 1D and KOMPSAT-1. Current high resolution sensors are described as "agile" and have a single telescope which can be oriented rapidly to enable fore and aft images to be obtained on a single pass.

1.3.5 Heights from stereoscopic data

As already noted, stereoscopic images can be used for measuring heights; the principle is shown in Figure 1.11. On two images, two points A and B of different heights will produce images at a_1 and b_1 on the left-hand photograph, and at a_2 and b_2 on the right-hand photograph. The separation of the images $a_1p_1 + a_2p_2$ $(= p_a)$ and $b_1p_1 + b_2p_2$ $(= p_b)$ is clearly proportional to the heights of A and B. The height (H) is related to the parallax by the equation

$$H = f \cdot \frac{B}{p_i} \tag{1.7}$$

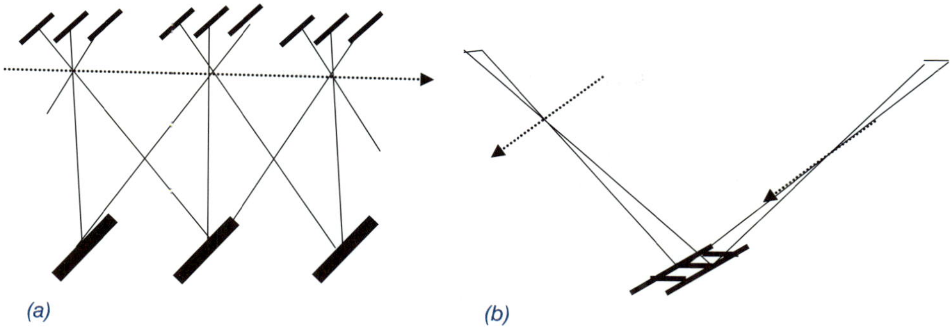

Figure 1.10 Methods of forming stereoscopic pairs: (a) along-track stereo; and (b) lateral stereo.

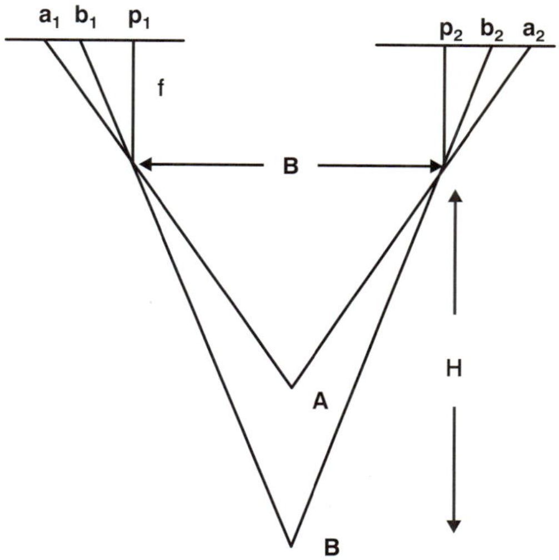

Figure 1.11 The principle of height determination.

where f is the principal distance, B is the base and p_i is the parallax of any point. These separations are called "parallaxes" or "disparities"; their measurement, as well as the rigorous determination of height, is discussed in Section 6.3.

Push-broom sensors are the most important system used for mapping from space. SPOT was the first of these and was designed as a mapping sensor. A push-broom sensor consists of a CCD (charge-coupled device) line in the focal plane of the mirror optics. The geometry of this system can be calibrated in the same way as a camera; however, the position and attitude of the sensor changes from line to line so that an image is made up of very narrow bands, each stable and with well-defined geometry but with a slightly different orientation to its neighbour. The mathematical model for a CCD line is simpler than that of a frame camera, but the model for a full-line scanner image is more complex. The movement of the sensor must be modelled and related to absolute reference systems. This is discussed in Section 6.3.

Stereoscopic overlap can be arranged either fore and aft or laterally so that 3D models can be formed. Displacement due to the relief of the ground is in the direction of view, and the amount of displacement is proportional to the incidence angle. If lateral overlap is used by tilting the sensor, the effect of relief will be large (see Figure 1.7) but a good base-to-height ratio will be obtained for height measurement.

1.3.6 Determination of position and orientation

In order to determine 3D information a sensor model is required; this is discussed in Chapter 6. A sensor model will determine the position and orientation of the sensor and allow 3D coordinates to be calculated from image coordinates. The required orientation information of a perspective image comprises the position of the sensor given in X, Y and Z coordinates in an appropriate reference system (which could be a Cartesian system with origin at the centre of the Earth) and the attitude of the sensor given as rotations about the coordinate system with a local origin. These six parameters of a perspective image or a push-broom line are known as the "elements of exterior orientation". They can be determined in a number of different ways, including:

- **Measurement on the platform or sensor (direct georeferencing)**. The position of satellites can be accurately determined. All high resolution optical Earth observation satellites are equipped with a global navigation satellite system (GNSS), such as GPS, star sensors and gyros. Star sensors lead to an attitude determination with lower-frequency rotations. Higher-frequency rotations have to be interpolated based on gyros relative to the base of the star sensor information. See Section 6.1.
- **Use of ground control points**. The position and attitude of the sensor can be calculated by relating the image to GCPs. This is relatively straightforward for photographic images but more complex when the image is formed over a period of time as with push-broom sensors. With a combination of sensor orientation and GCPs, the number of GCPs required can be reduced.

Chapter 6 covers the various models which are used with optical sensors from space.

1.3.7 Ground control

To determine the precise orientation and position of the sensor it is necessary to use GCPs. A GCP must be recognisable in the images and have known 3D coordinates in a suitable ground reference system. The number of points required depends on the method used and is discussed in the relevant sections. GCPs are also needed to check orientation carried out by direct georeferencing, or to validate orientation with control points – these may be called "independent checkpoints" (ICPs).

GCP coordinates may be obtained directly by survey measurement or from a map. In either case the coordinates will be given in a reference system: this may be geographical (latitude and longitude) or Cartesian (X, Y, Z), and may be globally based on the centre of the Earth or locally based on a national or regional projection. It is always important that the characteristics of the reference system be known and that all coordinates be given in the same system. Further information about reference systems and conversion between systems may be found in Iliffe and Lott (2008).

Direct survey measurements may come from surveys based on a local or national coordinate system, or they may come from a GNSS, for example GPS, which allows coordinates to be fixed directly from navigation satellites.

Maps should be used with caution for determining GCPs. Map data at scales of 1:25 000 and smaller is notoriously unreliable because of the many errors that may have accumulated in the map production and map digitising processes. These errors include survey errors (in some parts of the world published maps may be based on topographical sketches), drafting errors, generalisations, paper distortion and errors in digitising a paper document. It is always necessary to take into account the accuracy of the data when using ground control, and particularly important when using map data.

1.3.8 Accuracy

The accuracy of height determination from stereoscopic images can be calculated by differentiating Equation (1.7) with respect to p, which gives

$$\mathrm{d}H = \frac{H}{f} \cdot \frac{H}{B} \cdot \mathrm{d}p \tag{1.8}$$

Equation (1.8) is an expression which shows how a small change in parallax dp will give a small change in height dH. This equation can be used to give the precision with which height differences can be found. Precision is usually measured by the standard deviation of a set of measurements, or quantities derived from measurement. In this case the standard deviation of height determination σ_h can be found if the precision of observation of parallax σ_p is known, and the equation becomes

$$\sigma_h = \frac{H}{f} \cdot \frac{H}{B} \cdot \sigma_p \tag{1.9}$$

This expression shows that the precision with which height can be measured is related to the scale of the photograph (H/f), the base-to-height ratio (H/B) and the precision of measurement of parallax σ_p. For digital satellite images, the term $(H/f) . \sigma_p$ can also be expressed as a factor a multiplied by the GSD, leading to

$$\sigma_h = \frac{H}{B} . GSD . a \tag{1.10}$$

The factor a in Equation (1.10) depends on the image and object contrast and the nature of the land cover.

1.4 Optical imagery in context

Images from optical sensors are the most widely used data from space in the area of geomatics. However, other forms of data are becoming increasingly important, particularly from microwave sensors in the form of synthetic aperture radar (SAR). SAR sensors now obtain images with a 1 m resolution (TerraSAR-X) and airborne radar is highly competitive in terms of price and application (Figure 1.12). Interferometric SAR (IfSAR or InSAR) is highly effective for DEM generation and the Shuttle Radar Topography Mission (SRTM) has now achieved high quality global DEM coverage. Lidar data from airborne platforms is important, but satellite missions are limited by the power required for operating a laser system and by the ground coverage possible.

SAR is an active system which measures the distance (range) of a point from the sensor, and the backscatter received back at the sensor from a microwave pulse transmitted by the sensor. SAR only measures range and the azimuth direction, so in order to remove the effect of relief, the position of the sensor must be known and a DEM is required, as in the case of single optical images. A geocoded image (orthoimage) can then be generated. The geometry of SAR data is described in Schreier (1993).

Interferometric SAR allows the calculation of 3D data by using a pair of images taken with a short baseline and determining the phase difference in the two responses

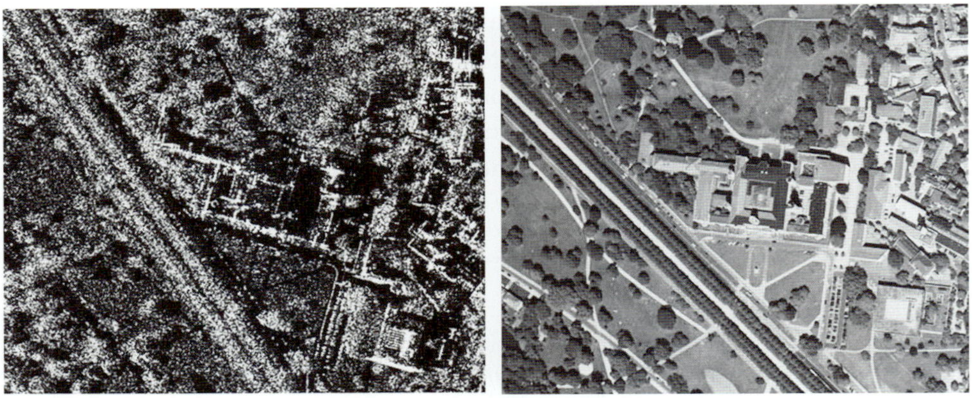

(a) (b)

Figure 1.12 SAR image compared with an optical image: (a) TerraSAR-X with 1 m GSD; and (b) aerial image.

from a single target, to form an interferogram which can be "unwrapped" to generate a terrain model. This is described in Gens and Van Genderen (1996). SRTM [NASA, 2011] is an important example of this technique, which produced a near-global DEM in an 11-day mission. A recent SAR system is TerraSAR-X [DLR, 2011], which was followed by TanDEM-X in 2010. TanDEM-X is a SAR interferometry mission which achieves image pairs by synchronised flying of TerraSAR-X and TanDEM-X to enable single-pass interferometry with base lines between 250 and 500 m. The target elevation accuracy for this system is ≤ 10 m absolute vertical accuracy and ≤ 2 m relative accuracy.

Table 1.3 sets out the characteristics of the optical systems SAR and lidar. The accuracy of different DEM generation systems relative to their cost is shown in Figure 1.13.

Table 1.3 A comparison of data from optical, radar and lidar sensors

	Optical	**Radar**	**Lidar**
Platform	Airborne/ spaceborne	Airborne/ spaceborne	Airborne
Radiation	Reflected sunlight	Own radiation	Own radiation
Spectrum	Visible/infrared	Microwave	Mostly infrared
Frequency	Multi-frequency	Multi-frequency	Generally single frequency, but dual frequency also used
Polarimetry	N/A	Polarimetric phase	N/A
Interferometry	N/A	Interferometric phase	N/A
Acquisition time	Day	Day/night	Day/night
Weather	Blocked by clouds	See through clouds	Blocked by clouds
Restitution	Position from orbit or GPS/INS and/ or GCPs and triangulation Object coordinates from stereo restitution	Position from orbit or GPS IfSAR required to obtain DSM GCPs also used	Position and orientation of sensor from GPS/ INS Object coordinates from range and azimuth Control areas also used
Products	Images 3D coordinates, including DSMs Orthoimages (geocoded images)	Images from SAR DSM from IfSAR Orthoimages (geo-coded images)	3D point cloud Intensity image

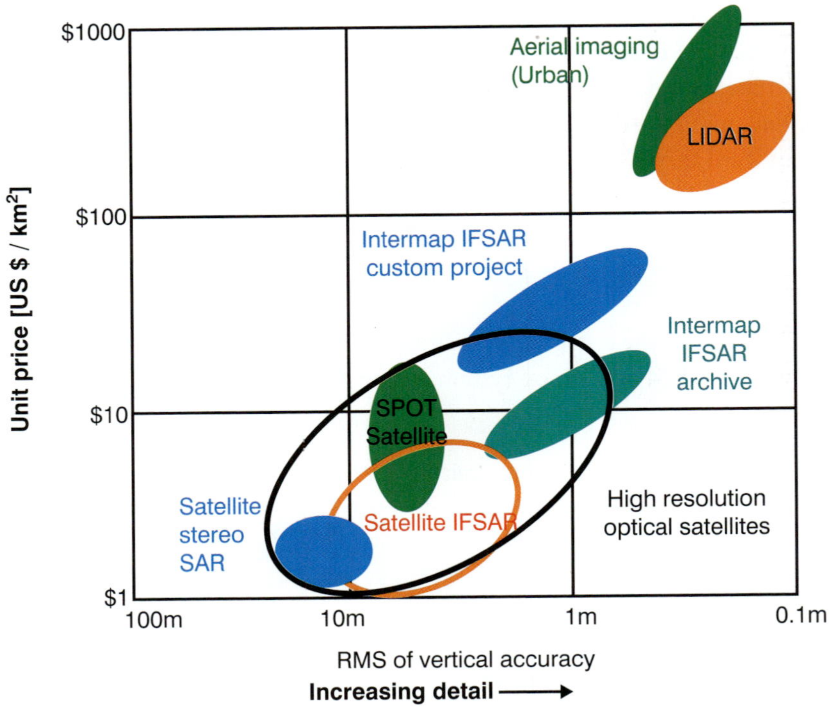

Figure 1.13 A comparison of sensors in terms of cost and accuracy. Credit: B Mercer, Intermap

1.5 Current systems

Table 1.4 sets out the sensors which will be covered in detail in this book, but also includes some sensors which will not be discussed further (only sensors which have been launched at the time of writing are included). Planned systems are discussed in Chapter 8.

Table 1.4 Summary of current sensors

Sensor	Agency Country launch date	Platform altitude (km)	GSD pan/MS (m)	Pointing capability (°)	Swath width at nadir (km)
Sensors with a GSD of greater than 5 m and up to 16 m					
HRV High-resolution visible	Spot Image France 1986, 1990, 1993, 1998	SPOT 1–4 832	10/20	sideways up to ±27	117 (i.e. 60 + 60 with 3 overlap)
MOMS-02	DLR Germany 1993	Space Shuttle 296	4.5/13.5	forward/nadir/aft –21.4/0/21.4	37/78
MOMS-2P	DLR Germany 1996	MIR space station 400	6/18	forward/nadir/aft –21.4/0/21.4	48/101
PAN (IRS-1C/1D) Panchromatic sensor	ISRO India 1995/1997	IRS-1C / IRS-1D 817	5.8	sideways up to ±26	70
ASTER Advanced Spaceborne Thermal Emission and Reflection Radiometer	NASA, METI USA, Japan 1999	Terra 705	VNIR: 15 stereo: 15 SWIR: 30 TIR: 90	nadir / aft 0°/27.7°	60
LISS-IV Linear imaging self-scanner-IV	ISRO India 2003	Resourcesat-1 817	5.8	sideways up to ±26	70

(Continued)

Table 1.4 (cont.) Summary of current sensors

Sensor	Agency Country launch date	Platform altitude (km)	GSD pan/MS (m)	Pointing capability (°)	Swath width at nadir (km)
AVNIR-2	JAXA Japan 2006	ALOS 699	10	free view to side up to 45	70
MSI Multi-spectral imager	RapidEye Germany 2008	RapidEye 5 satellites 530	6.5	sideways up to ±25	78
UoSAT-12	UK 1999	650	10/30	free view	10 m pan 30 m MS
Sensors with a GSD of 5m or less					
EROS A1	ImageSat Int. Cyprus (Israel) 2000	480	1.8	free view	12.6
HRS High-resolution stereoscopy	Spot Image France 2002	SPOT-5 830	5 × 10	forward/aft +20/– 20	120
HRG [a]	Spot Image France 2002	SPOT-5 830	5(2.5)/10	sideways up to ±27	60
FORMOSAT 2	NSPO China, Taiwan 2004	890	2/8	free view	24
PAN (Cartosat-1) Panchromatic sensor	ISRO India 2005	Cartosat-1 613	2.5	forward/aft 26/5 free view to side up to 23	27

TopSat Telescope	BNSC UK 2005	TopSat 686	2.8/5.6	free view	15/10
PRISM Panchromatic remote-sensing instrument for stereoscopy	JAXA Japan 2006	ALOS 699	2.5	forward/nadir/aft −24/0/+24 free view to side	70 35 (triplet stereo observations)
PAN (BJ-1) Panchromatic Imager	China NRSCC (CAST) 2005	BJ-1 Beijing 1 686	4/32	free view	24/640
CBERS-2B	CNSA / INPE China / Brazil 2007	778	2.4/20	free view	27/113
THEOS	GISTDA Thailand 2008	822	2/15	free view	22/90
AISat-2	Algeria 2008	680	2.5	up to 30 cross track	17.5
IKONOS-2	GeoEye Inc. USA 1999	681	0.82/3.2	free view	11.3
QuickBird	DigitalGlobe USA 2001	450	0.61/2.44 pan and MS only alternatively	free view	16.5
OrbView-3	GeoEye Inc. USA 2003	470	1/4	free view	8

(Continued)

Table 1.4 (cont.) Summary of current sensors

Sensor	Agency Country Launch	Platform altitude (km)	GSD pan/MS (m)	Pointing capability (°)	Swath width at nadir (km)
EROS B	ImageSat Int. N. Antilles (Israel) 2006	508	0.7/–	free view	7
Geoton-L1 Resurs-DK 1	Roscosmos (Roshy-dromet) Russia 2006	330–585	1/3 for h = 330 km	free view	30 for h = 330 km
MSC Multispectral camera	KARI South Korea 2006	KOMPSAT-2 685	1/4	sideways view up to ±30°	15
WorldView-1	DigitalGlobe USA 2007	494	0.45/–	free view	17.6
GeoEye-1	GeoEye Inc USA 2008	681	0.41/1.65	free view	15.2
PAN (Cartosat-2, 2A, 2B) Panchromatic camera	ISRO India	C artosat-2, 2007 Cartosat-2A, 2008 Cartosat-2B, 2010 631	0.82/–	free view	9.6
WorldView-2	DigitalGlobe USA 2009	770	0.45/1.8	free view	16.4

[a] Two HRG instruments on this mission can be processed to produce simulated imagery of 2.5 m.

References

DLR, 2011. http://www.dlr.de/eo/en/desktopdefault.aspx/tabid-5725/9296_read-15979 (accessed September 2011).

Forshaw, M.R.B, Haskell, A., Miller, P.F., Stanley, D.J. and Townshend, J.R.G., 1983. Spatial resolution of remotely sensed imagery – A review paper. *International Journal of Remote Sensing*, 4(3), 497–520.

GeoEye, 2011. http://www.geoeye.com/CorpSite/assets/docs/brochures/GeoSuite_Data_Sheet.pdf (accessed September 2011).

Gens, R. and Van Genderen, J.L., 1996. SAR interferometry – issues, techniques, applications. *International Journal of Remote Sensing*, 17(10), 1803–1835.

Iliffe, J.C. and Lott, R., 2008. *Datums and Map Projections*. Whittles Publishing, Dunbeath, UK.

ITRF, 2011. http://itrf.ensg.ign.fr (accessed September 2011).

Jacobsen, K., 2009. Potential of large format digital aerial cameras. Map World Forum, Hyderabad, GIS Development, http://www.ipi.uni-hannover.de/uploads/tx_tkpublikationen/DigitalCameras_Jacobsen.pdf (accessed October 2011).

Kratky, V., 1974. Geometric accuracy of ERTS. *Photogrammetric Engineering*, 40(2), 203–212.

Light, D., 1990. Characteristics of remote sensors for mapping and Earth science applications. *Photogrammetric Engineering and Remote Sensing*, 56(12), 1613–1623.

Light, D., 2004. A basis for estimating digital camera parameters. *Photogrammetric Engineering and Remote Sensing*, 70(3), 297–300.

Mikhail E., Bethel J. and McGlone J., 2001. *Introduction to Modern Photogrammetry*. John Wiley & Sons, New York, NY, USA.

Montenbruck, O. and Gill, E., 2000. *Satellite Orbits*, Springer Verlag, Berlin, Heidelberg and New York.

NASA, 2011. http://www2.jpl.nasa.gov/srtm (accessed September 2011).

Schreier, G. (Ed.), 1993. *SAR Geocoding: Data and Systems*. Wichmann, Karlsruhe, Germany.

Schreier, G. and Dech, S., 2005. High resolution Earth observation satellites and services in the next decade – a European perspective. *Acta Astronautica*, 57, 520–533.

Thomson, G., 2010. The practical effect of a diffraction-limited image for photogrammetry. *The Photogrammetric Record*, 25(130), 197–200.

Topan, H., Büyüksalih, G. and Jacobsen, K., 2005. Information contents of high resolution satellite images. EARSeL Workshop on 3D Remote Sensing, 9–11 June, Porto, Portugal, http://jeodezi.karaelmas.edu.tr/linkler/akademik/topan/yayin/08_ICHtop.pdf (accessed September 2011).

Xue, Y., Li, Y., Guang, J., Zhang, X. and Guo, J., 2008. Small satellite remote sensing and applications – history, current and future. *International Journal of Remote Sensing*, 29(15), 4339–4372.

History of Optical Sensors in Space | 2

Gottfried Konecny

2.1 History of space flight

The possibility of space flight was envisioned by dreamers. One of these was the author Jules Verne, the originator of science fiction literature. Born in Amiens, France, in 1828, Verne published his book *From Earth to the Moon* in 1865. Such literature encouraged others, such as Hermann Oberth (1894–1989), a Transylvanian Saxon, who, still a high school student, analysed the inconsistencies of Jules Verne's book with the laws of physics. He concluded that only rockets were capable of space travel, which he documented in his diploma thesis in physics. In 1929 he was contacted by Fritz Lang, a UFA movie producer from Babelsberg near Berlin, who worked on his film *Lady in the Moon* to give him some advice. (Universum Film Aktiengesellschaft, better known as "UFA" or "Ufa", was the principal film studio in Germany.) Later, his preoccupation with homemade rockets brought him an invitation to Peenemünde in 1941, where he met Wernher von Braun, who was engaged by the German military to develop the V-2 rocket.

Oberth was not alone as a prethinker for space travel. In Russia Mikhail K. Tikhonravov (1901–1974) worked with Sergei Korolev on rocket designs. Korolev's competitor in the Soviet Union, Valentin Petrovich Glushko, denounced Korolev in 1938 to Stalin, and Korolev was imprisoned until 1944, when Andrej Nikolajevich Tupolev succeeded in appointing Korolev as chief designer of the Soviet rocket development programme. In 1957 Korolev succeeded in launching the first Soviet intercontinental rocket. The launch of the first satellite, Sputnik, took place in 1957 (Figure 2.1), which Khrushchev accepted as a public relations success, although not officially a planned by-product. This opened the way to Yuri Gagarin's first manned spaceflight in 1961.

Wernher von Braun was engaged in rocket development from 1937 to 1945 in Peenemünde in Germany. After 1943, 3000 V-2 rockets were used during the war as weapons, reaching London and later Antwerp. When the Russian front moved forwards in Germany, the Peenemünde rocket team moved westward to be captured by the American forces in 1945. The US Army moved the team to the USA in 1946, when

Figure 2.1 Sputnik, the first satellite, launched in 1957. Credit: NSSDC Master catalogue, USA (2.9.1959)

Wernher von Braun was given the task of developing short-range missiles for the Army in Huntsville, Alabama.

At that time the US military goals were determined by the Cold War against the Soviet Union. The army, air force and navy had different (and even competing) strategies. The US Air Force first concentrated on the use of high altitude aircraft. The U-2, flying at an altitude of 25 km, permitted photographic overflights of the Soviet Union, China and Cuba without the danger of being intercepted by aircraft limited to lower altitudes. When a U-2 piloted by Powers was shot down by a Soviet rocket over Russia, all three US armed forces were convinced that they had to develop missiles. They were shocked by Korolev's success with intercontinental rockets and with Sputnik. Whereas the rocket launches of the air force and the navy failed, the US Army successfully launched the first American satellite, Explorer, on 31 January 1958.

Much earlier than this, Wernher von Braun had already become popular in the USA through Walt Disney's TV productions *Man in Space!* (1955), *Man on the Moon* (1955) and *Mars and Beyond* (1957), in which he discussed possible space flight goals. The events which took place in the Soviet Union and in the USA are discussed in depth in Matthew Brzezinski's book *Red Moon Rising* [Brzezinski, 2007].

When John F. Kennedy became US President in 1960, he made a pledge to put men on the moon within a decade and to bring them back to Earth. This became a boost for the computer industry and for analytical photogrammetry, to map landing sites on the moon for a safe landing, and to prepare for the exploration of the planets. NASA was established as an agency to implement the goals of space exploration, and Wernher von Braun became the first Director of the Marshall Space Flight Center in Huntsville, Alabama, where he was in charge of the Mercury, Gemini and Apollo programmes, leading to man's landing on the moon on 20 July 1969 in the Apollo 11 mission. During the space race between the Soviet Union and the USA, the military goals of observing parts of the Earth during the Cold War period were not abandoned.

2.2 CORONA programme

After the USA reached space capability in January 1958 and after the U-2 incident in May 1960, when Powers was shot down over Russia, the CIA began its military satellite reconnaissance programme under the name CORONA with the Key Hole satellites KH1 to KH4. The CORONA cameras are covered in more detail in Section 4.4.1. Altogether 144 CORONA satellites were launched between 1959 and 1972, with 102 satellites acquiring panoramic imagery on film, which at that time was the only appropriate high resolution medium. In 1959 a ground resolution of 7.5 m was achieved, with 2.75 m to follow in 1963 and 1.8 m in 1967. Many of the images were taken in stereo. The exposed film magazines were parachuted from the satellites and recovered by military aircraft. A schematic diagram of the panoramic camera used is given in Figure 2.2, and a CORONA image showing the Pentagon in Washington in Figure 2.3. In 1995 about 880000 of these images were declassified: they are now available at reproduction cost through the US National Archives and Records Administration (NARA).

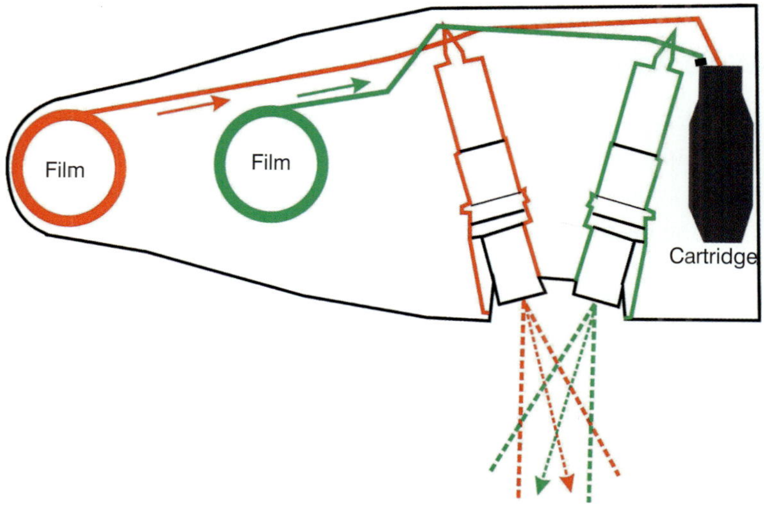

Figure 2.2 The CORONA panoramic camera. Credit: www.bernd-leitenberger.de/kh-1, source NRO

Figure 2.3 CORONA satellite image of the Pentagon, Washington, D.C. Credit: http://www.nro.gov/corona7.jpg (NRO)

2.3 Soviet Kosmos satellites

Similar to the developments in the USA, the USSR rocket development led to the first military satellite, Kosmos-1, with a photographic mission. Altogether 567 successful launches were carried out for military purposes between 1961 and 2001.

2.4 NASA's Apollo programme

From 1960 to 1972 NASA's Apollo programme in the USA bore witness to the difficulties of obtaining high resolution images from space, when high resolution digital sensors were not available. The purpose of the Apollo programme was the lunar landing of man. While the missions Mercury and Gemini opened the possibility for American manned space travel, other missions, such as Ranger 7 to 9 in 1964 and 1965, studied impacts of rockets on the lunar surface. Television cameras were used to obtain images from space before impact, which were transmitted to Earth. After the Surveyor mission confirmed the feasibility of landing a spacecraft on the moon, the task of mapping suitable landing sites was given to the Lunar Orbiter programme, which launched five satellites between 1966 and 1967.

The Lunar Orbiters were equipped with film cameras carrying Bimat film, which could be exposed and developed while orbiting the moon at an orbital height of 46 km above the lunar surface. After exposure and automatic development, the images were

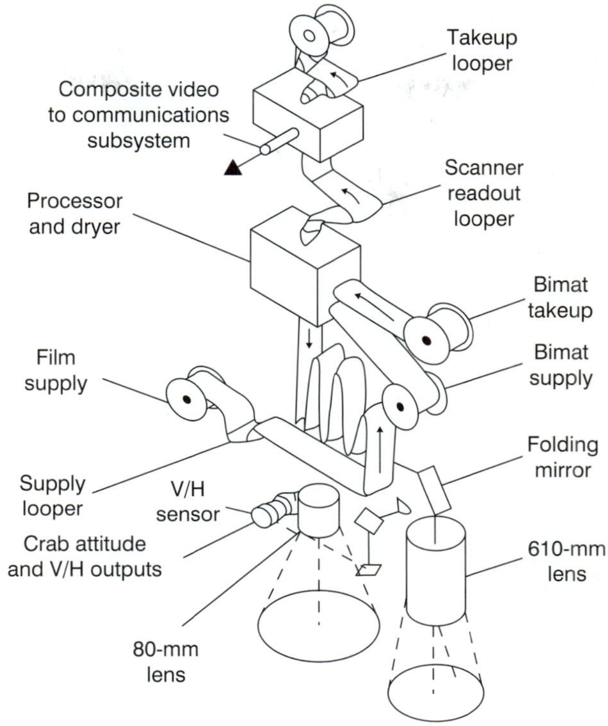

Figure 2.4 Lunar Orbiter camera. Credit: NASA/GSFC homepage

scanned in small strips called "framelets". They were then transmitted to Earth and reconstructed to full images again. The framelets contained pre-exposed fiducial marks on the film, and these permitted calibration of the highly distorted image parts in a bundle adjustment programme. Figure 2.4 shows a diagram of the camera; the framelets are clearly visible in Figure 2.5 in a Lunar Orbiter image of the Surveyor landing site [Konecny, 1968].

Despite the odd camera and imaging concept, the Lunar Orbiter programme became a great success. Twenty potential landing-site ellipses were mapped at high resolution from these images. The general location of these potential landing sites was first selected in relatively flat areas of the moon as seen on Earth-based telescopic images. The selected areas were overflown and imaged by the Lunar Orbiters. The high resolution images had a ground resolution of 4 m, and the medium resolution stereo coverage enabled generation of DEMs with 2 m relative heights. The mapping task for the landing sites was already accomplished from the first three Lunar Orbiter missions, so that missions four and five could be used to image the front and back portions of the moon from higher orbits. The last Apollo 17 mission was completed in December 1972 and a total of six astronaut landings took place on the lunar surface.

Figure 2.5 Lunar Orbiter image over the Surveyor landing site. Credit: NSSDC Image Catalog

The Soviet lunar programme with altogether 20 successful missions of Luna and Zond concentrated on unmanned exploration of the moon. Luna 3 was the first missile to impact the lunar surface in October 1959. Luna 9 succeeded in the first unmanned lunar landing in February 1966.

2.5 Earth observation missions

The civilian US Earth observation programme began with the weather satellite TIROS-1 (Television Infrared Observation Satellite 1), launched on 1 April 1960. It was equipped with a TV camera, transmitting some 22 000 images to Earth. The TIROS programme continued until 1965. Parallel to TIROS the weather research satellite programme NIMBUS took place from 1964 to 1978 with altogether seven satellites operating with a six-channel line scanner.

The follow-on satellites of the TIROS programme were ITOS and NOAA in 1970. In these the TV cameras were replaced by the operational Advanced Very High Resolution Radiometer (AVHRR), introducing digital sensor technology for Earth observation. NOAA satellites cover the Earth with 1 km ground sample distance (GSD) in five spectral channels. With two satellites, a twice-daily worldwide weather satellite watch became possible. The NOAA satellites were supplemented in the global coverage with (at that time) 5 km GSD from geostationary meteorological satellites GOES-1 and 2 (USA), Meteosat (ESA) and GMS (Japan), supplying images that were available every half hour. Figure 2.6 shows a TIROS-1 image, Figure 2.7 an NOAA image and Figure 2.8 a Meteosat meteorological satellite image.

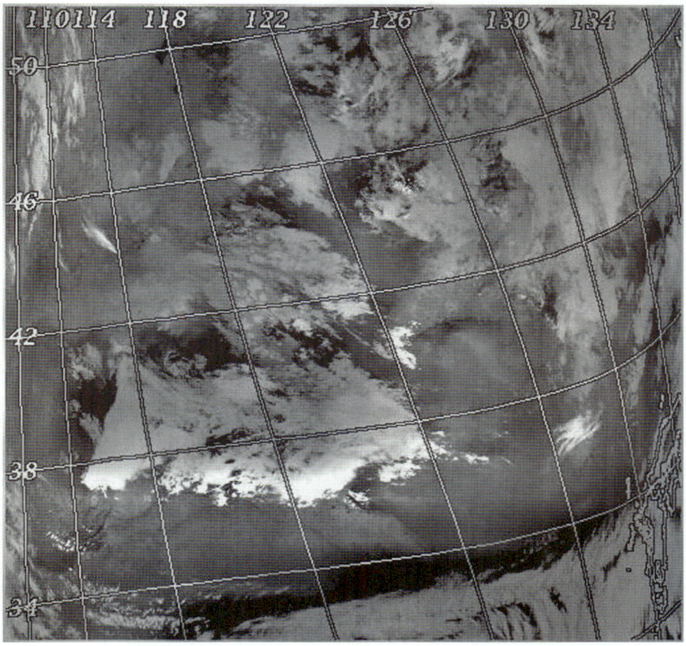

Figure 2.6 TIROS weather satellite image. Credit: http://earthobservatory.nasc.gov

Figure 2.7 NOAA satellite image of hurricane Katrina, 28 August 2005. Credit: www.noaa.gov

Figure 2.8 Meteosat satellite image over Europe. Credit: Eumetsat

In 1972 the first sun-synchronous Earth observation satellite, ERTS, later called Landsat-1, was launched. It had a TV sensor, which unfortunately failed, but the multispectral digital sensor, with four spectral channels at 80 m GSD, worked well. This launch established multispectral remote-sensing technology around the globe. Landsat-2 and 3 followed in 1975 and 1978, the latter with an added thermal band. Landsat-4, launched in 1982, and Landsat-5, in 1984, were equipped with a thematic mapper (GSD of 30 m in the visual and IR range, and a thermal band with 120 m GSD). Landsat-7, in 1999, added a 15 m GSD panchromatic band in the ETM (extended thematic mapper) sensor. Figure 2.9 shows a Landsat image taken over Mount Everest.

Landsat became the forerunner of subsequent Earth observation systems with push-broom array sensors, such as the French SPOT, launched in 1986, with other international developments following (India's IRS-1A in 1988, Japan's JERS-1 in 1992). SPOT permitted stereo imaging from different orbits by changing the inclination of the sensor. These satellite sensors were able to provide regional, and even global coverage, over an extended time, but they did not meet the high resolution requirements to be able to compete with aerial surveys.

2.6 Planetary missions

Digital technology for imaging was required for the first missions to explore other planets. For interplanetary distances, analogue image transmission did not reach the required energy levels, so digital grey level transmission (even for the early missions) was the only option. Mariner 4 acquired the first close-up images of Mars in 1965, from a distance of 10 000 km (Figure 2.10); other such missions were Mariner 7 in 1969 and Mariner 9 in 1971, before an unmanned landing on Mars became possible during the Viking mission in 1976. Mariner 10 sent images in bypassing Venus in 1973 and Mercury in 1974. The Voyager missions were designed to acquire images from the outer planets: Voyager 2 sent images from Jupiter in 1979 and from Saturn in 1981.

Figure 2.9 Landsat satellite image over Mount Everest. Credit: http://landsat.usgs.gov

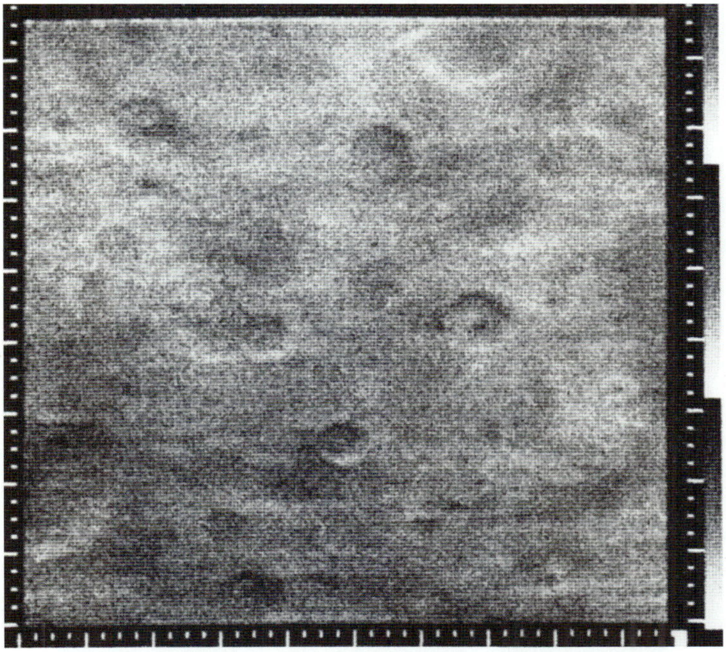

Figure 2.10 Mariner 4 image of the surface of Mars from a distance of 10 000 km. Credit: NASA, USA

2.7 Space Shuttle

Since 1981, the NASA Space Shuttle programme has not only allowed the sending of astronauts into space for various space experiments, but it also permitted utilisation of proven photographic high resolution sensors from the Spacelab module. Since 2009, 126 Space Shuttle missions have taken place, and mapping cameras have been part of the Shuttle payload on two occasions – the Metric Camera in 1983 and the Large Format Camera in 1984.

2.7.1 The German metric camera and the US large format camera

The first project of this type was the German metric camera, a modified Zeiss RMK 30/23 aerial photogrammetric mapping camera, which was accommodated in the first European Spacelab mission of 1983. The camera exposed about a thousand photographs in stereo, with a ground resolution of 10 m, and 10% of the land surface of the Earth was covered within a 10-day mission. An example is shown in Figure 2.11 [Konecny *et al.*, 1984].

Figure 2.11 Spacelab metric camera image over Tibet. Credit: www.dlr.de

Figure 2.12 The Itek large format camera (LFC). Credit: www.nasm.si.edu (Smithsonian National Air and Space Museum)

NASA launched the second mapping film camera, the American Itek LFC (large format camera), in 1984 (Figure 2.12). Exploitation of this classical, but effective, technology approach was halted because of the Challenger Shuttle disaster, which severely limited NASA's Shuttle flight resources. But this opened up the possibility for further US–German cooperation in launching the multispectral stereo scanner MOMS-02 on the Space Shuttle in 1993.

2.7.2 MOMS-02

MOMS-02 was a digital sensor incorporating several CCD (charge-coupled device) arrays imaging the Earth in a forward, nadir and aft panchromatic mode, with nominally 15 to 18 m GSD. Additional CCDs imaged the Earth in nadir mode in four spectral bands: a diagram is shown in Figure 2.13. The novel restitution geometry led to restitution software being developed at the German Aerospace Centre (DLR), in collaboration with German university photogrammetric institutes. The MOMS-02

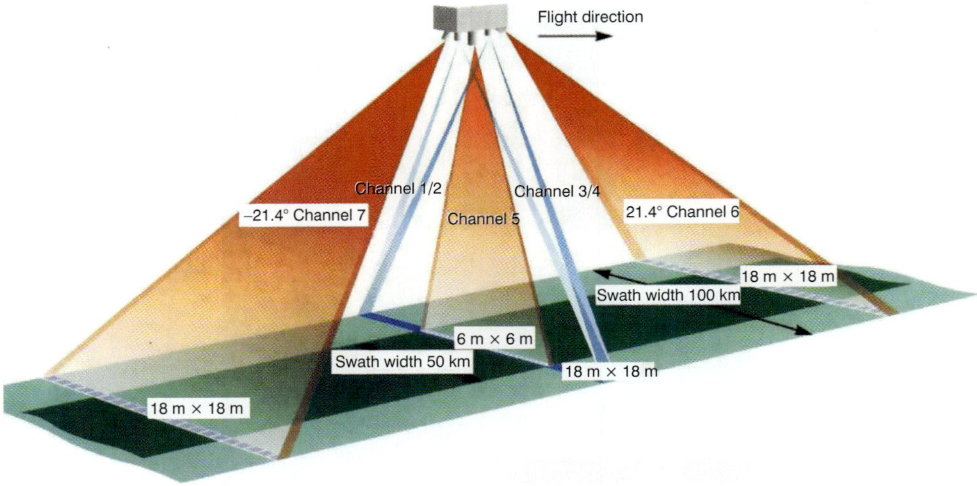

Figure 2.13 MOMS stereo scanner. Credit: www.dlr.de

stereo scanner may be considered as a forerunner of the current high resolution commercial satellites, such as IKONOS, QuickBird, WorldView, OrbView-3 and GeoEye-1.

MOMS-02 operated from Space Shuttle for a 10-day mission on Shuttle flight STS 55 from 26 April to 6 May 1993, at an orbital height of 296 km. It was subsequently installed on the Priroda module of the Russian MIR space station, where it operated from 1996 until the year 2001, covering 70% of the Earth's land surface with images [Bettac *et al.*, 1999].

2.8 East German and Russian photographic space technology

The international interest in Earth exploration from space was realised in the former socialist countries in the Interkosmos programme. The main sensor became the East German Zeiss-Jena film-based multispectral camera MKF-6, operated from the Russian Soyuz module starting in 1976. It had six objective lenses, with spectral filters in the visible and near-infrared range (Figure 2.14).

In 1987, at the ISPRS (International Society of Photogrammetry and Remote Sensing) symposium in Leipzig, the Soviet Union opened up the availability of Russian high resolution photographic imagery from the so-far classified cameras KFA 1000, MK4, KFA 3000 and KVR 1000. These are covered in more detail in Section 4.4.3.

Figure 2.14 The East German multispectral camera MKF-6. Credit: www.dlr.de

2.9 Development of current high resolution cameras

The US military CORONA programme (Section 2.2) from 1961 to 1972 reached higher resolutions, but it was a firm based panoramic camera system. Altogether 144 missions (102 successful) were flown before the program ended. As a follow on to the CORONA programme, the military Keyhole missions KH11 and KH12 began in 1976. KH11 was the first American spy satellite to employ digital image acquisition and transmission. It had reflective optics with a 2.3 m diameter mirror. The satellites, deployed in elliptical orbits, had a lifetime of about three years; a GSD of about 15 cm was claimed to have been achieved. KH12 is the follow-on mission, with improved optics (a 4 m diameter mirror), and so far six such satellites have been launched.

The key to reaching such extremely high resolutions from space was the development of time delay and integration (TDI) CCDs (see Sections 3.5 and 4.2). In TDI technology, not only is a linear CCD array used, but a matrix CCD is also introduced, consisting of several CCD lines. This is necessary since the velocity of a satellite passing over the surface of the Earth is about 7 km s^{-1}, which is equivalent to 7 m ms^{-1}. During that time, a chip having only a single CCD line cannot accumulate a sufficient number of photons for imaging at high resolution. Therefore the additional sensor lines of the chip-oriented perpendicular to the satellite motion pass on the photon counts received for each pixel, from line to line. As TDI chips of 85 lines in width have been industrially produced, and as there is a time period of 1 msec required to obtain a usable radiometric resolution for each pixel, a valid assumption is that the KH11 and KH12 satellites have used such TDI chips reaching a 15 cm GSD, provided this is possible subject to the diffraction limitations of the optical system, which in this case requires mirror optics of sizable dimensions. The KH11 and KH12 images must be very restricted in their area coverage.

The technology used in these military satellites has been licensed for use by US companies under certain conditions, and transferred to commercial high resolution satellites. Currently (2011) there are two main US companies operating and distributing high resolution satellite data: GeoEye and DigitalGlobe. These companies have developed from a number of mergers over recent years – companies such as Space Imaging and OrbImage have been incorporated into them. Figure 2.15 shows how GeoEye and DigitalGlobe have evolved.

GeoEye and DigitalGlobe count the US Defence Department as a main customer for images with a wide area coverage at a GSD of 0.5 to 1 m. Even though some commercial satellite sensors can image at 40 cm GSD, such images may not, by US rules, be made available internationally at this time. This indicates again that military technology, in developing instruments for satellites, has created a leap forward for civilian applications. However, civilian systems licensed in the USA are heavily dependent on contracts with the US government, and the future of these systems is linked to US government requirements and policies.

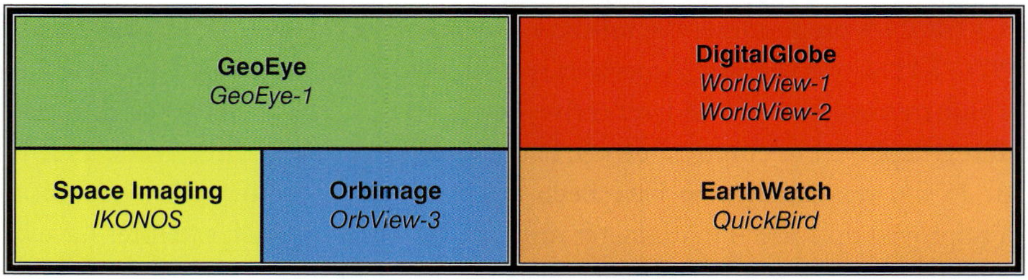

Figure 2.15 The evolution of the current main US commercial satellite operators(sensors are shown in italics).

References

Bettac, H.-D., Missling, K.-D., Panchenko, V., Tyan, T. and Voroncov, V., 1999. The project MOMS/ PRIRODA main results in 1996–1999. ISPRS Workshop Sensors and Mapping from Space, September 27–30, Hannover, Germany, http://www.ipi.uni-hannover.de/fileadmin/institut/pdf/bettac.pdf (accessed September 2011).

Brzezinski, M., 2007. *Red Moon Rising*. Times Books, Henry Holt & Co., New York, NY, USA.

Konecny, G., 1968. Some problems in the evaluation of lunar orbiter photography. *The Canadian Surveyor*, 22(4), 394–412.

Konecny, G., Reynolds, M. and Schroeder, M., 1984. Mapping from space: The metric camera experiment. *Science*, 225(4658), 167–169.

EUROPEAN SPACE IMAGING®
www.euspaceimaging.com
We've Got You Covered
Europe's leading VHR satellite data provider
WorldView-1: 50 cm panchromatic
WorldView-2: 50 cm 8-band multispectral
Ikonos: 1 m multispectral
Quickbird: 60 cm multispectral
agility
capability
flexibility
through local satellite tasking

3 Principles of High Resolution Optical Sensors

Rainer Sandau

3.1 Introduction

High resolution mapping systems have evolved since the development of the first civilian sensors and follow the trend towards smaller ground sample distances (GSDs). From the 80 m GSD of ERTS (renamed Landsat-1) in 1972, GSDs of less than 0.5 m for civil applications are now close to being achieved. The sensor system, the multispectral scanner (MSS) of Landsat-1, which was the first civil electronic mapping system, was based on a whisk-broom scanner. Modern systems are based on charge-coupled device (CCD) line arrays or matrix arrays using a push-broom scanner arrangement. Figure 3.1 shows the trend of decreasing GSD for civil Earth surface imagers using passive optical approaches. The growing number of spaceborne imaging systems in the last decade [see Kramer, 2002; Jacobsen, 2005] shows that more and more countries are dealing with spaceborne technology and that there is an increasing need for mapping systems for different applications [Konecny, 2003]. The trend of smaller GSDs was, and still is, supported by the improvements in diverse fields of technology such as optics, mechanics and materials, electronics, signal processing, communication and navigation.

A smaller GSD needs a larger focal length. The physics behind optical systems allows only a limited number of tricks to overcome the problems of large focal length optics in terms of volume and mass. The size of the focal plane depends on the detector system size and is part of the formula concerning optics, volume and mass. With decreasing GSDs, the requirements for pointing stability are increasing, and it is necessary to understand the requirements and restrictions. Higher resolution also means that it is necessary to deal with smaller amounts of energy coming from smaller ground pixels, which have to be registered in shorter integration time periods, according to the high satellite ground track velocities. In this context, important parameters for high resolution mapping, such as spatial resolution, radiometry, pointing accuracy and stability, must be understood. The parameters are derived from metrics based on a modulation

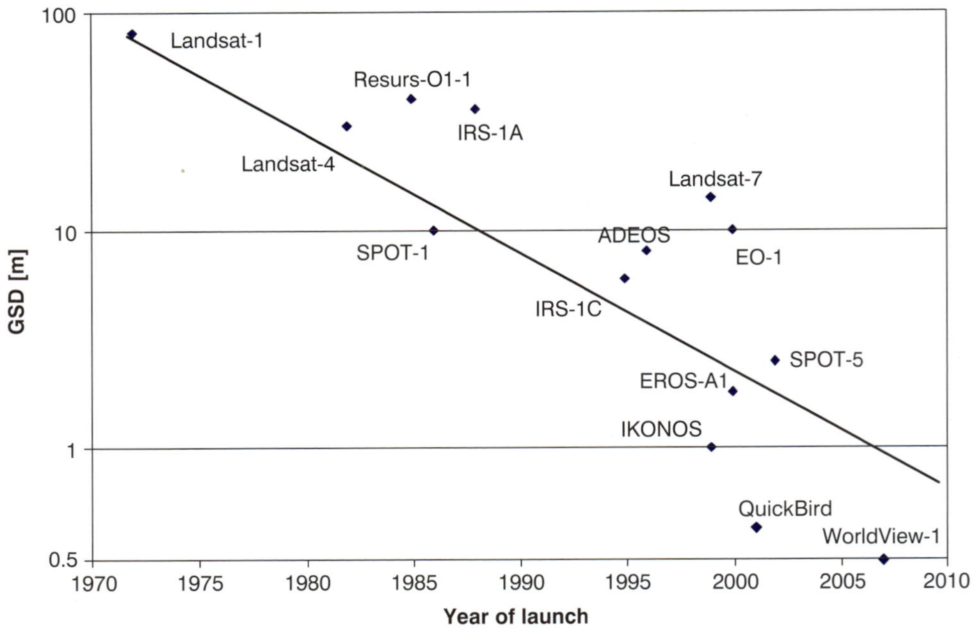

Figure 3.1 The trend of decreasing GSD as illustrated by some civil Earth surface imagers.

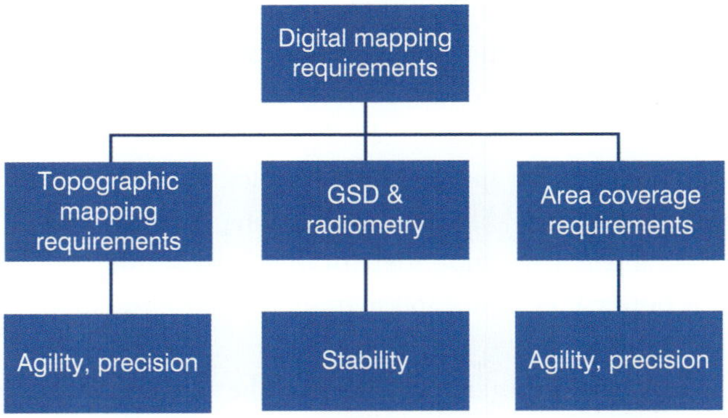

Figure 3.2 The requirements for digital mapping.

transfer function (MTF) [Sandau, 2004]. It is shown that the imager, as well as the spacecraft bus, needs to follow certain rules to make high resolution imaging possible.

If we think of topographic mapping, the dependency on the principles controlling the agility of the spacecraft or platform is another important parameter to consider in the context of imaging systems. Platform agility also needs to be considered in connection with multiple off-nadir angle observations. Figure 3.2 gives a general overview of the links between the requirements for digital spaceborne mapping and the set of conditions to be satisfied, such as pointing precision, pointing stability and agility of the spacecraft/platform. The importance of these conditions depends on the kind of requirements we are looking at: topographic mapping, GSD, radiometry and area coverage. All these issues will become transparent in the following sections.

The large amounts of data generated by high resolution imaging systems need to be stored and transmitted using high performance devices. Size, mass and power consumption of these devices increase with increasing data volumes and data rates. Data compression helps to reduce the data recording and transmission problems.

3.2 Topographic mapping concepts

3.2.1 Introduction

In this section the concepts of topographic mapping are discussed and it is assumed that the non-stereo acquisition modes are subsets of these concepts. Discussion will be restricted to the aspects relevant to the hardware designs and considerations, rather than the implications for photogrammetric data processing. There are several ways to perform topographic mapping, most of them making use of line or matrix technologies.

3.2.2 Across-track stereo

This concept (see Figure 3.3) is used by the push-broom imaging systems of SPOT 1–4. The area to be mapped topographically is imaged in a side view mode in orbit number K. The second view angle necessary for stereo computation is provided in orbit $K+N\,(N \geq 1)$, when the same area is mapped in the side view mode from the other side. Clearly, the time gap between the two orbits can cause some problems for the stereo reconstruction: there are, for instance, different illumination conditions or even seasonal changes to be taken into account.

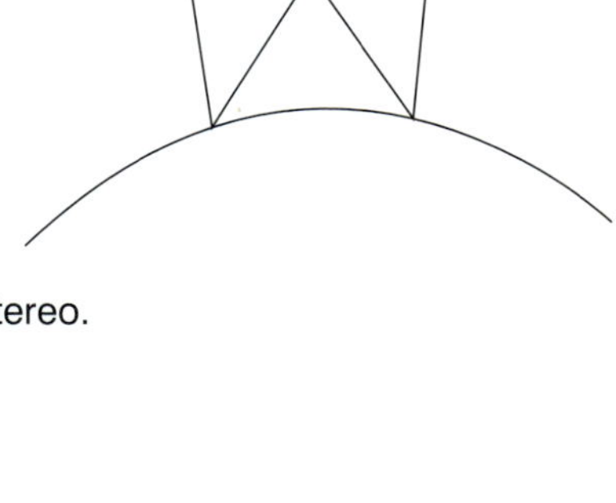

Figure 3.3 Across-track stereo.

3.2.3 Along-track stereo

Along-track stereo avoids the illumination problems associated with across-track stereo imaging: the areas to be mapped are imaged during one pass from different directions – for instance forward and backward. The illumination conditions can be considered to be constant during the small time interval between the imaging sequences for the different view angles (less than two minutes; see also Section 3.4). There are different concepts in use, ranging from a one-camera concept to a three-camera concept; additionally, the cameras can be equipped with line arrays or matrix arrays. Furthermore, multiple camera systems can be substituted by one-camera systems with multiple line arrays accommodated on the single focal plane (imaging plane) behind a single lens cone. The distances between the line arrays on the focal plane are chosen according to the intended stereo angles of base-to-height ratios (B:H ratios; see Section 1.3.4).

Figure 3.4 shows the working principle of the one-camera concept using a forward/backward slewing feature. The first image of the stereo pair is generated in the forward-looking mode. The area can be mapped using a push-broom scanner equipped with a line array detector, or by using a matrix camera generating consequential images according to the strip length. The second image is then generated in the backward-looking mode. This concept provides a variable stereo angle according to the stereo baseline used for the image pair. IKONOS, QuickBird and WorldView are examples of push-broom camera systems, and the Turkish BilSat-1 is an example of a matrix camera system.

With a two-camera system on board a satellite, both image views can be generated simultaneously (see Figure 3.5). Due to the movement of the satellite, the area imaged first by the forward-looking camera is then imaged a second time when the second sensor system covers the same area. The stereo angle is fixed because of the fixed viewing angles of the two-sensor system. Actually, this concept is implemented using CCD-line

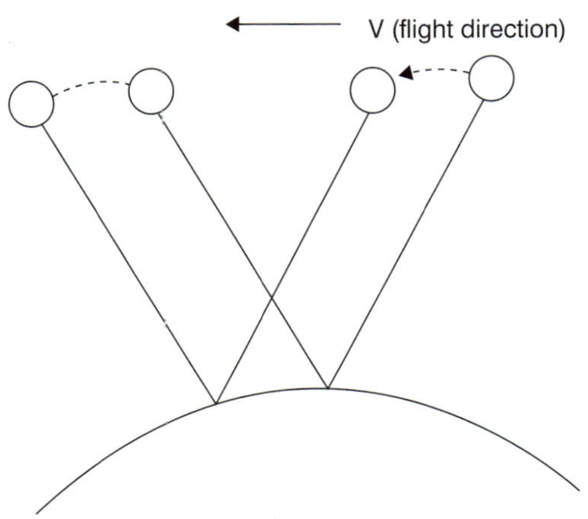

Figure 3.4 Along-track stereo, using the slewing feature of the satellite.

sensor systems. The CCD lines are placed either in two different cameras, for instance two HRS sensors on SPOT-5, or on the focal plane of a single camera, for instance OPS on JERS-1, or ASTER on TERRA.

With three sensors, (Figure 3.6) imaging with three different view angles, there are some advantages (which may not apply simultaneously):

- Stereo imaging is more robust using the information coming from three view directions.
- Platform or satellite disturbances during the imaging sequences can be eliminated by applying linear regression algorithms.
- View angles are fixed so that it is possible to select stereo pairs with stereo angles best suited to the actual terrain type (see Table 3.1) [Sandau, 2010].

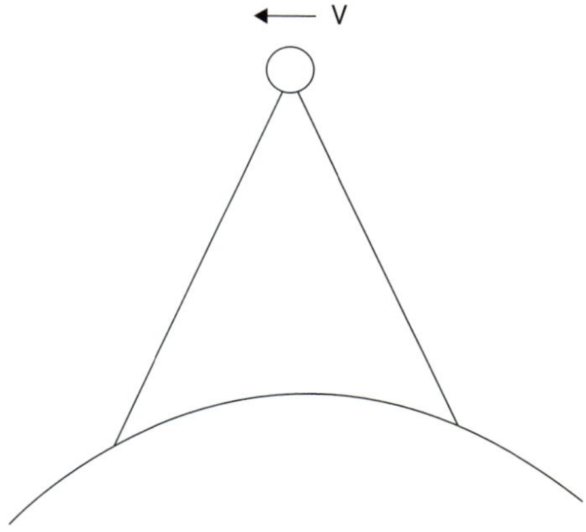

Figure 3.5 Along-track stereo with two sensors.

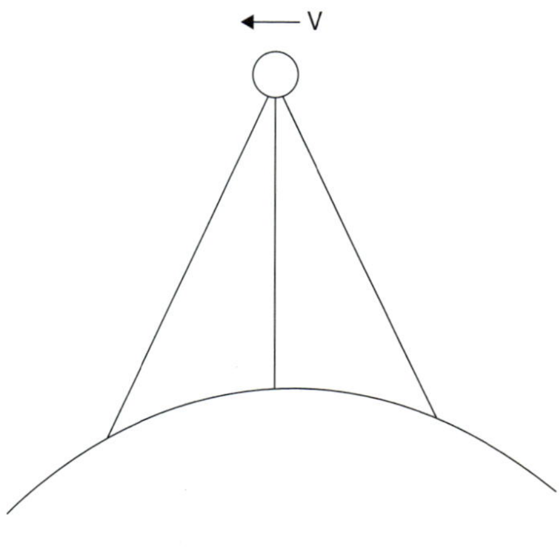

Figure 3.6 Along-track stereo with three sensors: three lines in the focal plane or three single-line cameras (e.g. MOMS-02).

Table 3.1 Stereo angles for various applications [Sandau, 2005]

Topographic applications	Stereo angle (°)
Flat terrain and increased height accuracy	30–60
Hilly terrain	20–40
Mountainous areas	10–25
Object extraction applications	
Natural landscape	30–50
Suburban areas	20–40
Urban areas	10–25
Woodland	10–25

3.3 Important parameters for spaceborne mapping systems

3.3.1 Spatial resolution

The most important features required to make the relationships between the different components of a high resolution optical spaceborne mapping system transparent are described in this section. Some of the major features that influence the image quality from the spatial resolution point of view are considered.

A very effective way of describing the image quality is the use of a modulation transfer function (MTF). With the MTF approach, all the MTF components of linear systems (or quasi-linear systems) influencing the image quality, which may be based on different physical effects (e.g. optics and CCD detector), are combined in order to create the system MTF. The resulting point spread function (PSF) of the system is then computed by applying the inverse Fourier transform (IFT). Using the MTF approach, by working in the frequency domain rather than the spatial domain, we avoid the need to deal with complicated and time-consuming convolution algorithms. But, of course, the MTF approach is not applicable if non-linear components are to be considered.

The MTF of the spatial resolution, MTF_{SR}, is composed of two main elements: the "static", or instrumental, component and the "dynamic", or platform stability part.

$$MTF_{SR} = MTF_{instrument} \, . \, MTF_{PS} \tag{3.1a}$$

Since $MTF_{instrument}$ is mainly influenced by the optics and the detector element, Equation (3.1a) can be rewritten as

$$MTF_{SR} = MTF_{Optics} \, . \, MTF_{D} \, . \, MTF_{PS} \tag{3.1b}$$

where "SR" denotes spatial resolution, "D" denotes detector and "PS" denotes platform stability.

The MTF of the platform stability, MTF_{PS}, is the subject of Section 3.3. MTF_{Optics} includes the diffraction part as well as the aberration part. For most operating spaceborne systems, the optical system may be considered to be near-diffraction limited and in focus. For a diffraction-limited circular aperture of the optics, the MTF_{Optics} can be described by the equation

$$MTF_{Optics}(k_x, k_y) = \left| \frac{2}{\pi} \left[\arccos(\lambda \frac{f}{D} k) - \lambda \frac{f}{D} k \sqrt{1 - (\lambda \frac{f}{D} k)^2} \right] \right| \quad for \ \lambda \frac{f}{D} k \leq 1 \qquad (3.2)$$

where f is the focal length, D is the aperture diameter of the optics, λ is the average wavelength of the radiation, and $k = \sqrt{k_x^2 + k_y^2}$ (spatial frequencies). This diffraction causes a diffraction disk, or "Airy disk". The diameter d of the Airy disk caused by diffraction is derived from PSF_{Optics} [Morris *et al.*, 2008] and given by

$$d = 2.44 \,.\, \lambda \,.\, \frac{f}{D} = 2.44 \,.\, \lambda \,.\, F \qquad (3.3)$$

where $F = f/d$ is the F-number. The diameter d is one of the important parameters which can be related to the detector pixel size x. If x is larger than d the system is detector limited and the resolution is determined by the detector. Otherwise, the optics determines the spatial resolution. Figure 3.7 shows the borderline for an average wavelength of $\lambda = 0.55$ µm. The optics design should be near to the borderline on the optics-limited side in order to obtain maximum energy for the detector while avoiding too large an aliasing effect. For state-of-the-art CCD detectors with a pitch of 7 µm, an *f*/5.2 optics would satisfy this condition.

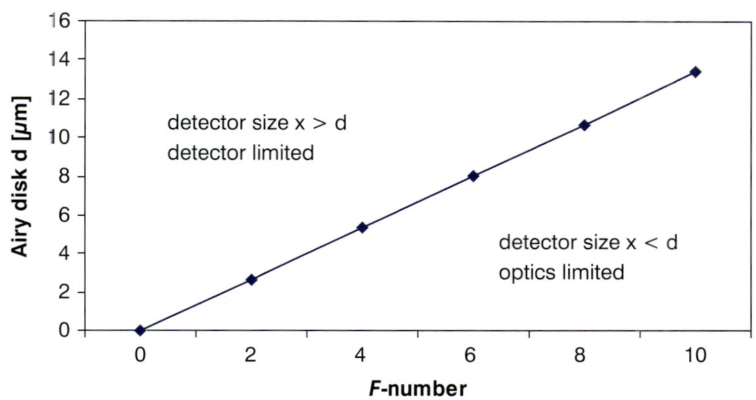

Figure 3.7 Airy disk parameter d as a function of the *F*-number *F* ($\lambda = 0.55$ µm).

MTF_D of a rectangular detector element with disk parameter d is described by the sinc functions

$$MTF_D\,(k_x) = \text{sinc}(\pi \,.\, x \,.\, k_x) \,.\, \text{sinc}(\pi \,.\, y \,.\, k_y)$$
$$= \frac{\sin(\pi \,.\, x \,.\, k_x)}{\pi \,.\, x \,.\, k_x} \cdot \frac{\text{sinc}(\pi \,.\, \text{y} \,.\, k_y)}{\pi \,.\, y \,.\, k} \tag{3.4}$$

For simplicity we assume quadratic pixels here and describe the subsystem using a one-dimensional (1D) formula when the extension to the two-dimensional (2D) case is obvious or can be easily done. MTF_D is depicted in Figure 3.8, in which the abscissa is the spatial frequency normalised to the detector specific value $k = k_x / k_{x,max}$. With a detector pitch of 10 μm, the maximum observable spatial frequency is 100 lp mm^{-1} (line pairs per millimetre). The maximum spatial frequency that can be reconstructed from the imagery is the Nyquist frequency $k_{Nyquist}$. According to the Nyquist Theorem, a frequency must be sampled at least twice in order to reconstruct it correctly and consequently $k_{Nyquist} = k_{max} / 2$. For our example, this yields $k_{Nyquist} = 50$ lp mm^{-1} (Figure 3.9). If the detector pitch x is given in μm, $k_{Nyquist}$ can be computed using the formula $k_{Nyquist} = 1000 / (2 \,.\, x)$.

3.3.2 Radiometric aspects

The number of photoelectrons generated in a solid-state camera is

$$n_{pc} = \frac{A_D \,.\, T_{Optics} \,.\, t_{int}}{4F^2} \int_{\lambda_1}^{\lambda_2} R_d(\lambda) L(d\lambda) d\lambda \tag{3.5}$$

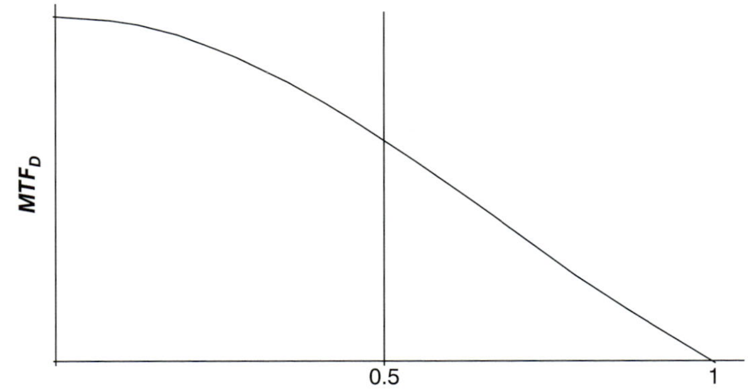

Figure 3.8 MTF$_D$ as a function of the normalized spatial frequency. k = 0.5 is the Nyquist frequency $k_{Nyquist}$.

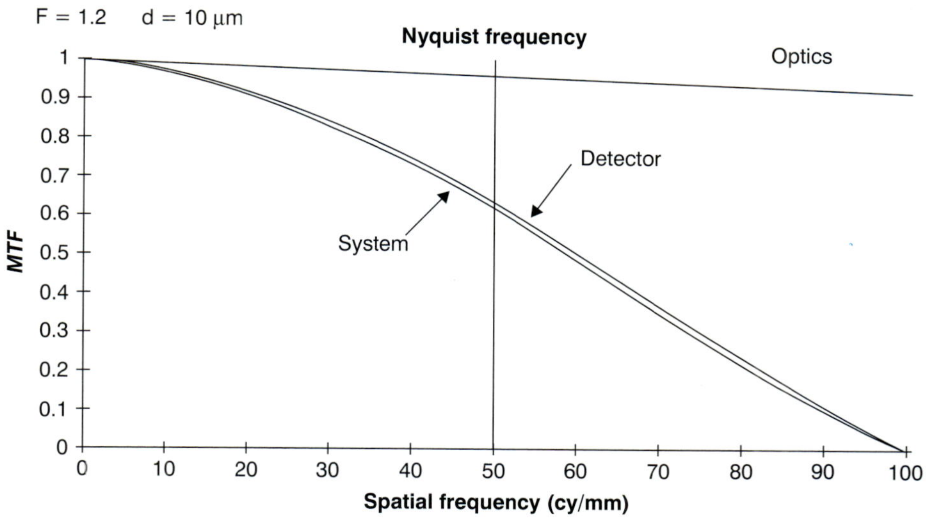

Figure 3.9 System MTF composed of MTF_{Optics} and MTF_D: the relationship between a CCD pixel size of x = 10 μm behind an *f*/1.2 optics. In this example, MTF_{Optics} could be neglected.

where A_D is the detector area, T_{Optics} is the transmission of the optics, t_{int} is the integration time, F is the F-number, R_d is the detector responsivity and L is the radiation flux with $t_{int} < t_{dwell}$. Once the detector is selected, A_D and R_d are given. L is also given, as well as F and T_{Optics} when the optics is selected or designed taking into account the technological and/or the mission constraints. $\Delta\lambda$ is fixed in most cases, so the only real variable part is the integration time t_{int}.

For a satellite in low Earth orbit (LEO), the satellite ground track velocity is about 7 km s^{-1}. In other words, the dwell time is 1 ms for a GSD of 7 m. For the actual technological practicalities, a GSD of 10 m delivers sufficiently adequate energy during the dwell time to allow direct imaging in the panchromatic channel as well as in broadband multispectral channels, with a satisfactory signal and signal-to-noise ratio (SNR) for mapping. Actually, the RapidEye cameras use this approach, achieving a GSD of 6.5 m. But if the GSD is reduced from 10 m to (for instance) 1 m, the consequences severely affect the system design – the dwell time would be reduced by factor of 10

$$t_{dwell}\,(1\text{m}))\,/\,t_{dwell}(10\text{m}) = 1/10 \tag{3.6}$$

Even more severe is the influence on the pixel instantaneous field of view (*IFOV*).

$$IFOV(1\text{m}))\,/\,IFOV(10\text{m}) \triangleq 1/100 \tag{3.7}$$

When both of these aspects are taken into account, a reduction of the GSD by a factor of ten causes a time-related and geometry-related decrease in energy at the detector by a factor of about a thousand. There are two possible ways of overcoming this problem

- Use time delay and integration (TDI) technology (see Section 3.4) with N stages to increase the signal N-fold and improve the SNR by a factor of $\sqrt{N}$ (e.g. this technique is used in the IKONOS and QuickBird missions).
- Use the so-called "slow-down mode" to decrease the ground track velocity of the line projection on the surface with respect to the satellite velocity in order to obtain the necessary increased dwell time t_{dwell} (e.g. this technique is used in the EROS-A1 mission).

3.3.3 Pointing accuracy and stability

There are many ongoing activities for developing and testing instruments, actuators and algorithms to control pointing with high precision. The accuracies obtained are of the order somewhere between arcseconds and fractions of degrees. For mapping of the Earth's surface, deviations from the necessary accuracy can be corrected using precise GCPs. WorldView-1 may serve as an example for a state-of-the-art mapping system [Spilker, 1997]. With a GSD of 0.5 m at nadir, the following parameters are given

Pointing accuracy	< 500 m at image start and stop
Geolocation accuracy	6.5 m (may be improved with GCPs)

The pointing stability is of more importance in order to maintain the GSD and the image quality. The MTF of the platform stability, MTF_{PS}, has three major components

$$MTF_{PS} = MTF_{LM} \, . \, MTF_J \, . \, MTF_{sin} \tag{3.8}$$

where "PS" denotes platform stability, "J" denotes jitter, "sin" denotes sinusoidal vibration and "LM" denotes linear motion. The MTF degradation due to linear motion of the satellite is

$$MTF_{LM}(k_x) = sinc(\pi \, . \, a_{LM} \, . \, k_x) \tag{3.9}$$

where k_x is the spatial frequency and a_{LM} is the distance moved by the target edge across the detector pixel. MTF_{LM} only affects the MTF in the direction of the motion. The distance a_{LM} is equal to $v \, . \, \Delta t$ ($\Delta t = t_{int}$). In many cases, Δt is close to the dwell time and MTF_{LM} is approximately equal to MTF_D. Figure 3.10 shows the influence of a_{LM} on

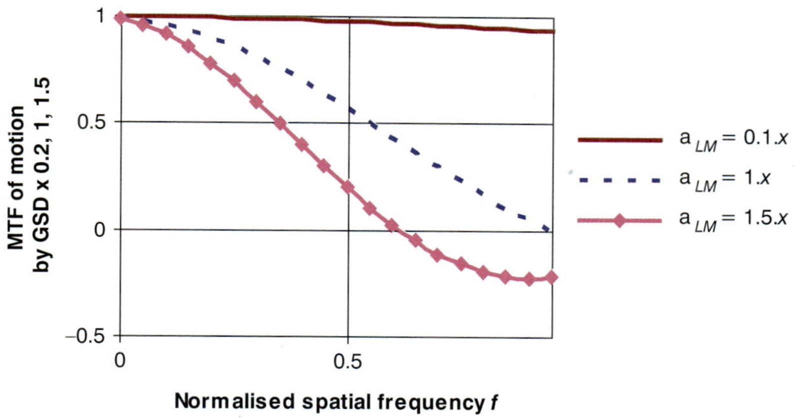

Figure 3.10 Influence of a_{LM} on MTF_{LM}.

MTF_{LM}, where x is the detector element size; when $a_{LM} = 1$, MTF_{LM} equals MTF_D. The abscissa shows the spatial frequency normalised to the system dependent maximum value $k = k_x / k_{x,max}$. For instance, with a detector pitch of 6.5 µm the spatial frequency of 150 cyc mm^{-1} equals $f_x = 1$. As a rule of thumb, when the linear motion causes an image shift of less than about 20% of the detector size, the effect on system performance is minimal.

Jitter is another problem: for MTF_J (jitter or random motion) it is assumed that the jitter is a superposition of different high-frequency motions so that the central limit theorem of statistics can be applied. The theorem states that the superposition of many random movements with different statistical distributions results in a Gaussian distribution characterised by a single parameter – the standard deviation σ

$$MTF(k_J) = \exp\left(-2\pi^2\sigma_J^2 k_J^2\right) \tag{3.10}$$

where σ_J is the standard deviation of random displacement and $k_J = \sqrt{k_{J,x}^2 + k_{J,}y}$ is the spectral frequency.

Figure 3.11 shows the influence of σ on MTF_J for $\sigma = 0.1 \times x$ and $\sigma = 1 \times x$ (x is the detector element size). For comparison reasons, MTF_D of the detector is depicted as well. With $\sigma = 1$, $MTF_{instrument}$ is disturbed in such a way that the intended GSD is not achievable. As a rule of thumb, when σ_J is less than about 20% of the detector size x, system, performance is only minimally affected. Attitude control systems for pointing accuracies and stabilisation to support high resolution functions on micro satellites are currently under development. In some cases, vibrational disturbances may also be avoided by simply switching off the active control functions during the relatively short imaging phase.

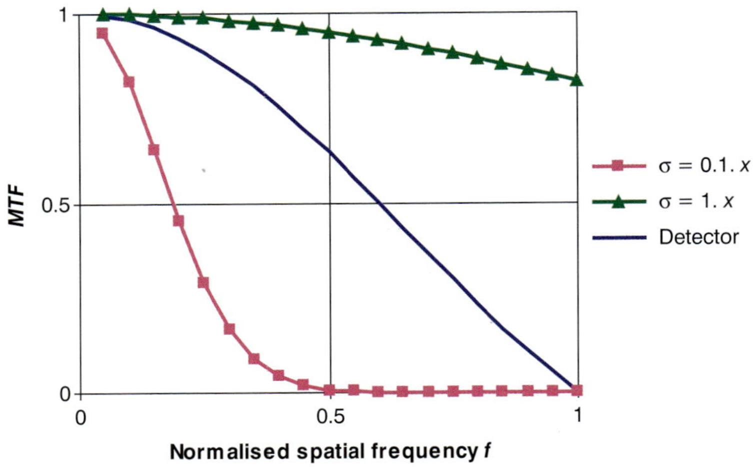

Figure 3.11 Influence of σ on MTF_J. MTF_D is shown for comparison.

A sinusoidal platform vibration will be considered for systems with larger integration times (lower ground resolution) using

$$MTF_{sin}(k_x) = J_0(2\pi a_{sin} k_x)$$
$$t_{int} >> \frac{2\pi}{\omega} \qquad (3.11)$$

where J_0 is the Bessel function of order zero and a_{sin} is the amplitude of the sine wave. Figure 3.12 shows the influence of the amplitude a_{sin} of a sinusoidal platform vibration. Again, with $a_{sin} = x$ the camera platform combination is not able to provide the intended GSD. As a rule of thumb, when the amplitude a_{sin} is less than 10% of the pixel size, the effect on system performance is minimal.

Figure 3.13 shows the degradation of MTF_D (with detector size x) due to the influence of MTF_{PF} when the rule of thumb parameters are applied (MTF_{LM} with $a_{LM} = 0.2 \times x$; MTF_J with $\sigma = 0.1 \times x$; and MTF_{sin} with $a_{sin} = 0.1 \times x$) for all three components of MTF_{PF}. The resulting MTF_{pf1} equals MTF_{SR} with neglected MTF_{Optics} ($MTF_{Optics} = 1$).

From an orbit altitude of 600 km, a GSD of 1 m equals an IFOV of 1.7 mrad, or approximately 0.3 of an arcsec. During the dwell time, the drift should be less than 20% of the IFOV, resulting in a drift rate of about 2.4 mrad s^{-1}, or 8 arcmin s^{-1}, in order to stay in the limit for minimal degradation of the MTF due to drift effects. When using the TDI principle to improve the SNR, for a 96-step TDI the tolerable drift rate even becomes 25 µ s^{-1}, or about 5 arcsec s^{-1}! These requirements are also applicable to systems using the slow-down mode to increase the dwell time. One possibility to fulfil such high pointing stability requirements is to use cascaded pointing systems, i.e. rough

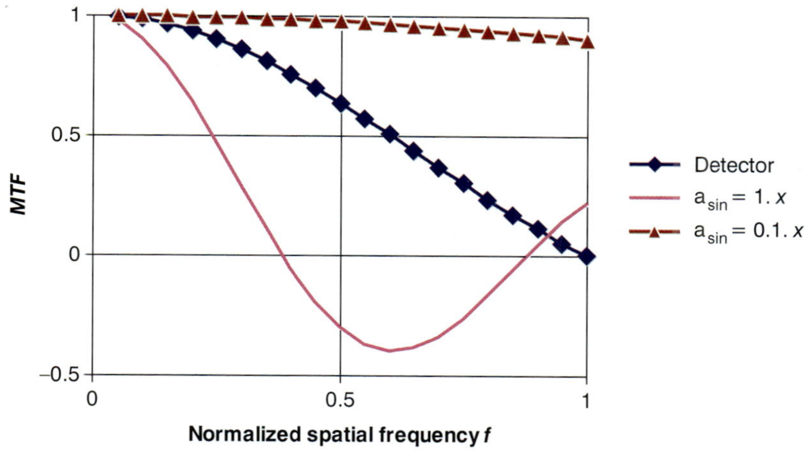

Figure 3.12 Influence of the amplitude a_{sin} of a sinusoidal vibration on MTF_{sin}. MTF_D is shown for comparison.

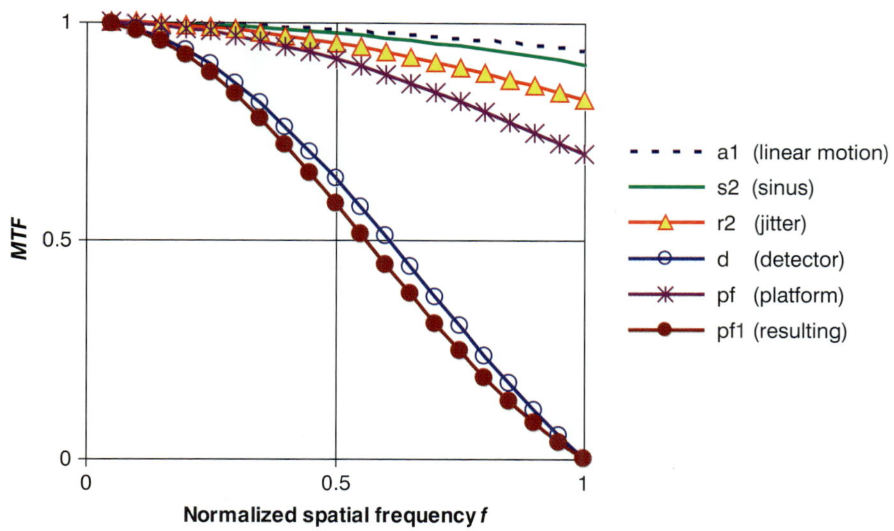

Figure 3.13 Degradation of MTF_D (MTF_d) due to MTF_{PS} (MTF_{pf}), consisting of MTF_{LM} (a_{LM} = 0.2 . x), MTF_J (σ = 0.1 . x), MTF_{sin} (a_{sin} = 0.1 . x). The resulting MTF is MTF_{pf1}.

pointing by one system and fine tuning of the pointing by an additional system. If we look into, for instance, the QuickBird parameters [Krause, 2005], the pointing stability is given with < 10 mrad s^{-1} per axis.

3.3.4 Agility

From Figure 1.10 of Chapter 1 we can see that agility is an important feature of mapping systems when we apply certain stereo-imaging concepts or strive to cover as much as possible of an area on as many as possible different ground sites, according to customer request or the actual cloud coverage conditions. Figure 3.14 shows some collection scenarios which can be realised with, for instance, WorldView-1 [Krause, 2005]. They enable different imaging tasks to be performed during one pass over a certain area.

To give an idea of the agility parameters such as acceleration and retargeting rate, the stereo-imaging mode may serve as an example. For stereo imaging with a one-sensor concept, the agility depends on the B:H ratio. For example, from a 600 km orbit and a GSD = 1 m, assuming a B:H = 1 and using a 10 000 × 10 000 matrix camera, we achieve a stereo pair of about 60 km × 60 km ground area. The camera view angle needs to be retargeted by about 100° within 74 s, resulting in an average turning rate of about 1.4° s^{-1}. If it is necessary to cover other areas as well, a high retargeting capability is necessary, including the acceleration capability at the start of retargeting and before it stops. WorldView-1 provides good performance: for instance, for an acceleration of 2.5° s^{-2}, the rate is 4.5° s^{-1}.

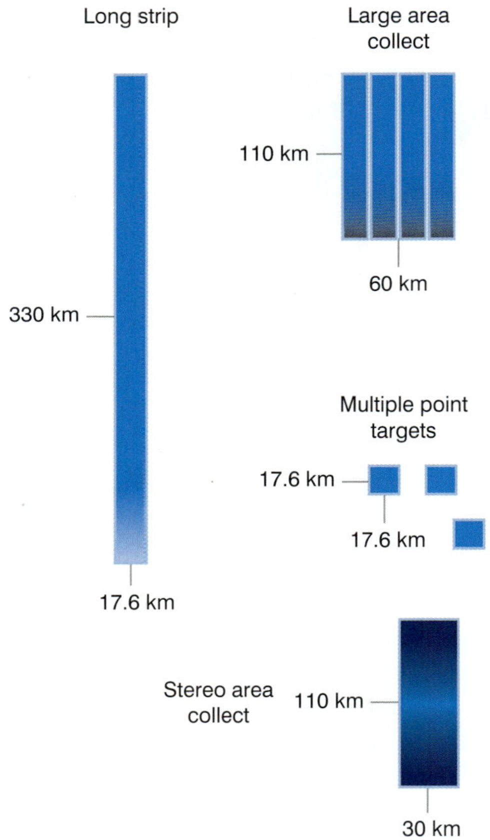

Figure 3.14 Example of image collection scenarios with a one-sensor concept (see Section 3.2) [DigitalGlobe, 2009]. http://worldview2.digitalglobe.com/docs/Constellation_Brochure.pdf).

3.4 Sensor optics

This section deals with design considerations for the optics and the focal plane. The focal length of high resolution spaceborne sensors is determined by the laws of physics. The basic parameters of an optics are focal length f, the ratio $f/D = F$ (the F-number) and the optical quality in terms of lp mm^{-1}. The focal length can be calculated using

$$f = \frac{H_{Orbit}}{GSD} \cdot x \tag{3.12}$$

where x is the detector pitch. Figure 3.15 illustrates this formula for an orbit height $H = 600$ km and $GSD = 1$ m.

For a pixel pitch of 10 μm (the distance between pixel centres, approximately the same as pixel size), the focal length f is 6 m (see Equation (3.12) and Figure 3.15). According to Equation (3.15) and Figure 3.7, the Nyquist frequency is $k_{Nyquist} = 50$ lp mm^{-1}. From Equation (3.3) and Figure 3.7 we can extract the F-number, $F = 8$. Using that, the aperture of the optics is determined to be $D = 0.75$ m. Table 3.2 shows the optics parameters for an imager giving a $GSD = 1$ m from $H = 600$ km for two detector pitch values: $x = 10$ μm and $x = 1$ μm.

Obviously, from the focal length point of view, the development of detectors of size 1 μm should be pursued. However, it is likely there will be problems in providing the

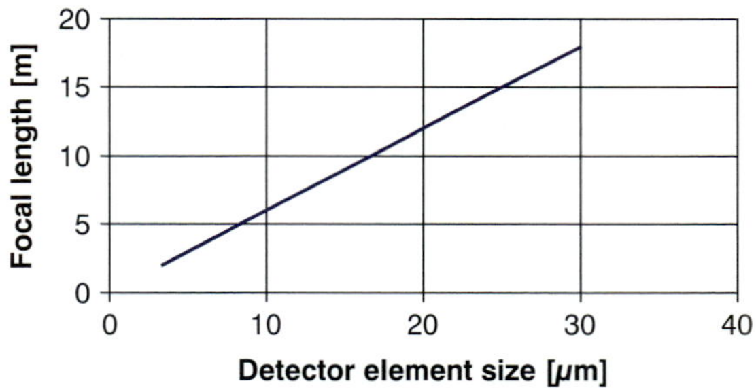

Figure 3.15 Relationship between detector element size X and necessary focal length f for a given GSD = 1 m from an orbit altitude of 600 km.

Table 3.2 Optics parameters for H = 600 km, GSD = 1 m, for given detector pitch x

Detector pitch (μm)	Focal length (m)	F-number	Aperture (m)	Nyquist frequency (lp mm^{-1})
10	6	7.5	0.8	50
1	0.6	0.75	0.8	500

Nyquist frequency of 500 lp mm^{-1}. Also, the energy problem needs to be considered. According to Equation (3.5) the number of photoelectrons generated from the photons hitting the detector is proportional to the detector area. Decreasing this area from 10 μm × 10 μm to 1 μm × 1 μm will enlarge our radiometry problem (see Section 3.2) by a factor of 100, leading to 100 times more TDI steps or increasing the slow-down feature by a factor of 100. Most of the cameras designed use detectors with a pitch value between 1 and 10 μm.

Generally, there are two basic concepts in designing optics for space applications: refractive and reflective. Refractive systems are mainly used when the resolution power requirement is moderate, for instance $f_{Nyquist}$ < 100 1p/mm. This is connected to pixel pitches $x \geq 5$ μm and F-numbers of the order of four and above. Nowadays, using the newest lens production and test equipment, powerful lightweight and low-volume refractive telescopes based on aspheric lenses can be designed. Refractive telescopes are used in many spaceborne systems, especially when higher resolution power is required. There are several systems in practical use. A trade-off has to be made between the advantages and the disadvantages of the different concepts, by taking into account the manufacturer's capabilities in terms of complexity, as well as the costs [Cutter *et al.*, 2008]. In this context the major concepts shown in Figure 3.16 are characterised as follows:

- **Newtonian-derived telescopes** have simple mirror shapes and a very small obscuration. However, they are long and need a complex relay lens.
- **Cassegrain-derived telescopes** have simple spherical mirrors that are easy to produce and align, but large diameters are needed. The secondary mirror is the bottleneck in terms of the diffraction limit. These types of telescope are used, for instance, on the SSTL-150 and SSTL-300 satellites, providing GSDs of 4 m and 2.5 m, respectively.
- A **three-mirror anastigmatic (TMA) telescope** has no aperture obscuration, and allows compact and lightweight designs. But it has expensive off-axis mirrors and requires a complex alignment. This type of telescope has been used in space on the small satellite TopSat, achieving a GSD of 2.8 m.
- The **Korsch telescope** design allows a small hole in the primary mirror, but is complex with high mounting sensitivities. The Rutherford Appleton Laboratory (RAL) and MacDonald Dettwiler and Associates (MDA) have developed the next-generation optical systems for small satellites, achieving a GSD of 1 m (RALCam-4) and 0.5 m (RALCam-5), based on a Korsch/TMA-derived concept [Morris *et al.*, 2008].

There are many possible design variations based on the refractive or reflective concepts, as well as hybrid versions, leading to sophisticated catadioptric designs. The information given here is intended to give some insight into the possibilities offered and the constraints faced when dealing with high resolution optical spaceborne sensors.

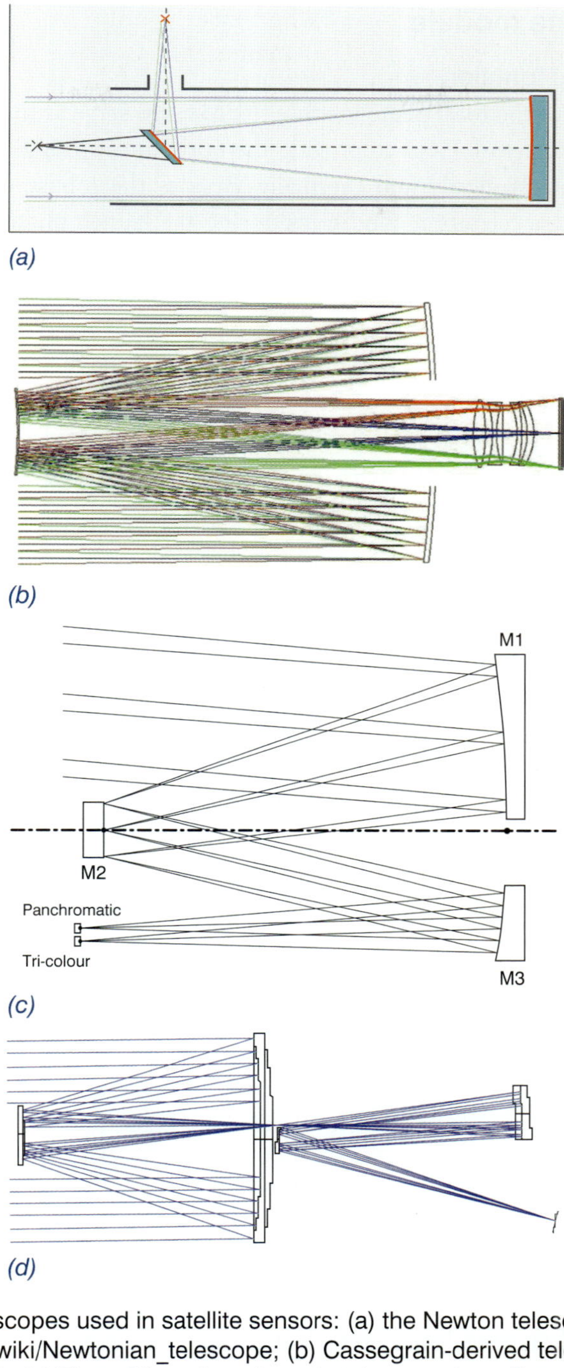

Figure 3.16 Telescopes used in satellite sensors: (a) the Newton telescope from http://en.wikipedia.org/wiki/Newtonian_telescope; (b) Cassegrain-derived telescopes from Cutter *et al*. (2008); (c) Three-Mirror Anastigmatic telescope from Cutter *et al*. (2008); and (d) Korsch telescope from Cutter *et al*. (2008).

3.5 Focal plane module

3.5.1 Mechanical function

The basic function of the focal plane module (FPM) is to serve as the mounting structure to locate the detector(s) and to provide the optimum working conditions for the detector(s), both mechanically and thermally. Besides the robustness to survive the harsh vibration and acceleration conditions during the launch phase, the focal plane's flatness is a key point in order to ensure a constant focal length constraint over the entire image field. Furthermore, the thermal extension coefficients of the base material of both the detector and the FPM need to match in order to avoid thermal/mechanical stress in phases when the thermal control is not active. If the focus control is not provided by components of the optics, it needs to be implemented in the focal plate structure. The front-end electronics with the analogue read-out circuitry is also attached to the FPM, which allows short distances between detector outputs and read-out to provide optimum noise behaviour.

3.5.2 Thermal function

The detectors, accommodated on the FPM, are sensitive to thermal variations. In particular, the dark signal of CCD devices is strongly dependent on the temperature: the dark signal is doubled for a temperature increase of about 7 degrees Kelvin. This has a strong influence on the dynamic range and the SNR of the CCD and, if not corrected, on the accuracy of the data. So, the material and structure of the FPM base need to support the homogeneous temperature distribution within the detector areas. To minimise the energy consumption for thermal control, the front-end electronics, which is mechanically integrated in the FPM, is often thermally decoupled from the FPM.

3.5.3 Detectors

It was mentioned earlier that linear or matrix arrays are the normal technologies used for recording light intensity. These two different sensor technologies coexist in the field of spaceborne mapping systems. For instance, EarlyBird uses a matrix approach, whereas IKONOS, QuickBird and WorldView use a linear array approach. In the case of very high resolution mapping, the choice has consequences on the operation mode – whether to use TDI or slow-down mode to increase the dwell time. With the matrix approach, a snapshot of the area to be imaged is made and the integration time must not exceed the dwell time, i.e. $t_{int} \leq t_{dwell}$, according to the conditions described in the linear motion discussion of Section 3.3 (see also Figure 3.10). With that in mind, only the slow-down mode is applicable. For push-broom sensors, however, both options are available – TDI or slow-down mode. If the system has to also provide multispectral information, the push-broom technology has advantages. With additional *N* line arrays accommodated in the FPM, *N* spectral channels (in-field separation of colour channels) can be built up and complex beam-splitting focal plane assemblies avoided, as in the case of the matrix technology (with the exception of using Bayer arrays).

There are possibilities and restrictions with linear arrays. Linear arrays are produced with large pixel sizes representing the potential *GSD/S* ratio, where *S* is the swath width. If the available number of pixels does not fulfil the *GSD/S* ratio, there are two possibilities to extend the number of pixels (see Figure 3.17). Possibility one, the buttoned configuration (Figure 3,17b), has the disadvantage of having a gap in the middle caused by technological needs. Possibility two, the overlap configuration (Figure 3.17c), avoids the gap but shows a certain divergence angle between the left and the right array. These assemblies can also be implemented using TDI devices.

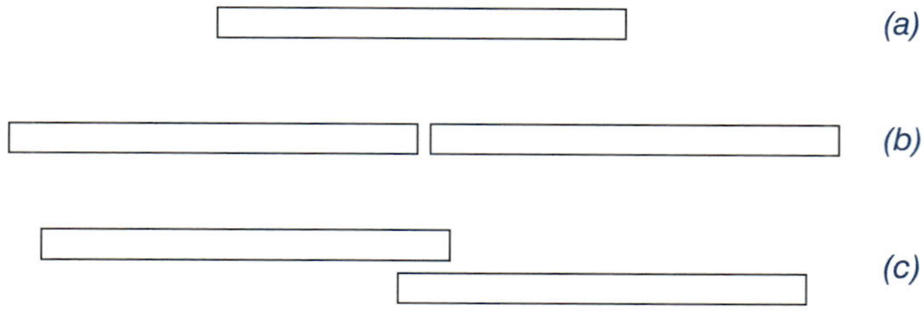

Figure 3.17 Linear array extension: (a) no extension; (b) buttoned extension; and (c) overlap configuration.

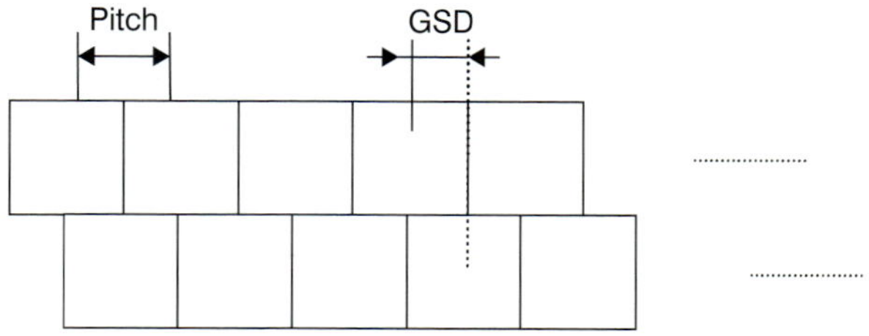

Figure 3.18 Staggered linear detector array configuration.

But there is yet another possible way to decrease the GSD (increase the ground resolution), keeping the swath width *S* – this is the "staggered configuration" (see Figure 3.18). The staggered configuration has major impacts on the system design. The volume and mass of the optics are strongly dependent not only on the focal length and the aperture, but also on the image field size determined by the detector extension. Having the same number of ground sample points, the image field size is halved with respect to the configuration of Figure 3.17(b) or (c). Additionally, the focal length is halved. In summary

- detector line length is halved;
- image field area is reduced to one-quarter;

- focal length is halved;
- optics needs to be of high quality for twice as many line pairs per millimetre with respect to the line pairs per millimetre necessary for the pixel size.

Or, looking at it the other way around, with optics as shown in Table 3.2 for GSD = 1 m and x = 10 µm (f = 6 m, D = 0.8 m), a GSD of 0.5 m could be achieved provided the optics quality is good for $f_{Nyquist}$ = 100 1p/mm. But note that a TDI version of a staggered configuration is very complicated or even impossible with state-of-the-art technology. There are two ways to implement the staggered arrays as shown in Figure 3.19 – supermode and hypermode. The names "supermode" and "hypermode" were created by the SPOT-5 development team.

A CCD line is placed at the focal plane of a telescope in the linear mode. It provides a GSD equivalent to d_1 in both CCD-line and flight direction. This mode is used in most of the push-broom imagers. In the supermode, as depicted in Figure 3.19, a double CCD linear array is placed at the focal plane of a telescope, which yields two shifted classical images. When the shift of the staggered line (line 2) is 0.5 pixels along both CCD and flight directions, we obtain the sample pattern of the supermode with a minimum GSD equivalent to $d_s = d_1/\sqrt{2}$. This mode is used, for example, in the SPOT-5 mission cameras HRS [Boissin *et al.*, 2002]. Physically, CCD line 2 is shifted 0.5 pixels in the CCD-line direction and 3.5 pixels in the flight direction.

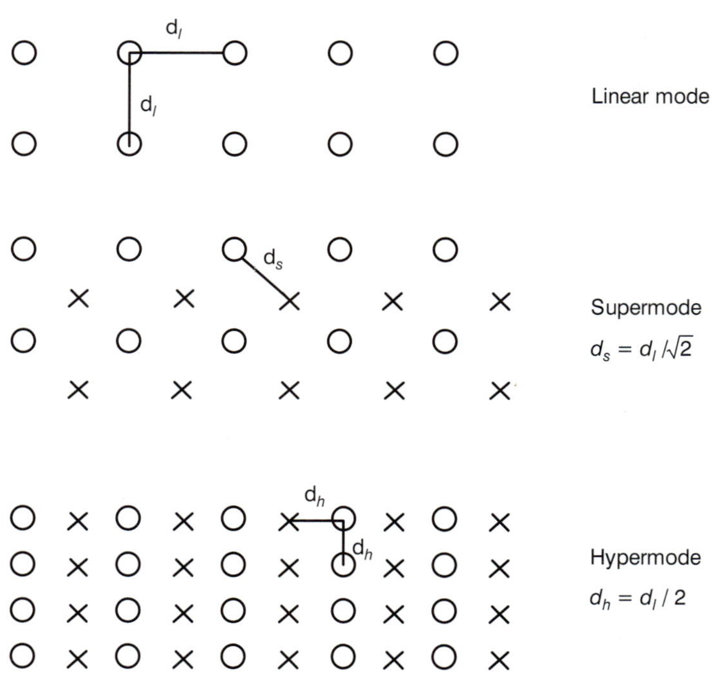

Figure 3.19 Different sampling modes with staggered arrays: o=linear array 1; × = linear array 2 (staggered).

In the hypermode, when the shift of the staggered line is 0.5 pixels along the CCD-line direction and both CCD lines are sampled twice during the dwell time of 1 pixel, we achieve the ground sample pattern of the hypermode with a GSD equivalent to $d_h = d_1/2$. This mode is used, for instance, in the BIRD mission IR cameras [Brieß *et al.*, 2003] and also in the airborne digital sensor ADS40 of Leica [Sandau, 2010].

3.6 Data recording and transmission

3.6.1 Introduction

Sensor data rates are very important parameters for imaging systems. This key parameter, in relation to the available downlink capacity, is crucial for choosing one of the three options:

1. Image and real-time downlink.
2. Store and dump (image and store, downlink later).
3. Image and store, downlink other data at the same time.

Conversely, having requirements according to the three options above, the storage and downlink capacity has to be designed appropriately. The input data rates of the data recorder are related to the sensor data rates, and the output data rates are related to the downlink data rates, respectively. The data rate problems can be reduced using data compression methods, either in real time before the data recorder, which decreases also the data recorder capacity requirements, or before the downlink, which allows various on-board data processing tasks. These tasks include on-board data analysis, compression and encryption.

3.6.2 Sensor data rates and volumes

Sensor data rates can be treated in the same manner for both linear and matrix sensor technologies because both sensors have exactly the same pixel dwell time under the same circumstances (*GSD*, focal length, detector size, orbit height). Sensor data rates depend on

- the number of pixels across flight direction N of a linear or matrix array;
- radiometric resolution M in number of bits per pixel;
- ground track velocity (depends on the orbit altitude);
- ground sample distance (*GSD*).

The ground track velocity V_{ground} and *GSD* determine the effective dwell time

$$t_{dwell} = \frac{GSD}{V_{ground}} \tag{3.13}$$

The sensor data rate DR is determined in bits per second (bps) by

$$DR = \frac{N \,.\, M}{t_{dwell}} \tag{3.14}$$

The data volume DV, in bits, generated within the imaging time t_{image} is

$$DV = DR \,.\, t_{image} \tag{3.15}$$

As an example, with H_{orbit} = 600 km ($V_{ground} \approx 7$ km s^{-1}), a CCD line with N_y= 12 Kpixels, a radiometric resolution M = 8 bit, and a quadratic image N_x = N_y = 12 Kpixels, the imaging time is $t_{image} \approx 15 \,.\, 2^{10} \,.\, t_{dwell}$. These assumptions are equivalent to using a matrix with 12 000 × 12 000 pixels. The results for different *GSD*s are given in Table 3.3.

Using the staggered array configuration, the *GSD* = 10 m of the linear array can be decreased to *GSD* = 7 m in the supermode and to *GSD* = 5 m in the hypermode. The related values for *DR* and *DV* are shown in Table 3.4.

In the case of WorldView-1, the maximum contiguous area imaged (mono) is 60 km × 110 km. With a GSD of about 40 cm this yields 330 Gb for 8-bit radiometric resolution. This corresponds well with the numbers given in the data sheet: 2199 Gb for on-board storage and 331 Gb per orbit collection. With this buffer capacity, the operation options 1 and 3 mentioned above are supported.

Table 3.3 Data rates and volumes for *GSD* = 10 m and *GSD* = 1 m (H_{orbit} = 600 km, 8-bit radiometric resolution)

GSD **(m)**	***DR*** **(Mbps)**	***DV*** **(MB)**
10	70	150
1	700	150

Table 3.4 Sensor data rates and volumes for different sample modes (H_{orbit} = 600 km, 8-bit radiometric resolution)

Mode	***GSD*** **(m)**	***DR*** **(Mbps)**	***DV*** **(MB)**
Linear	10	70	150
Super	7	140	300
Hyper	5	280	600

In addition to the panchromatic channel, most of the satellite systems used for mapping have multispectral channels with somewhat coarser GSDs, and they have to be considered as well when analysing the sensor data rates and volumes. For instance, three multispectral bands of the SPOT or TopSat missions, with GSDs coarser by a factor of two, will result in an additional data rate and volume totalling 75% of that generated by the panchromatic channel. On the other hand, the satellites of the RapidEye constellation provide five multispectral bands with GSD = 6.5 m each, without a separate panchromatic channel.

3.6.3 Data transmission

A primary consideration for the design of satellite communication systems is the efficient use of transmitted power and channel bandwidth, as space communication links are both power and bandwidth limited. For space to Earth data transmission, a frequency spectrum of about 20 MHz in S band, 375 MHz in X band and 1500 MHz in Ka band is allotted. Each has different requirements for meeting the most important objectives of satellite communications – the power and bandwidth/spectrum efficiency, combined with a robust bit error rate (BER) performance [Spilker, 1997].

At the moment, X band transmitters are in common use for spaceborne high resolution mapping systems. The data coming from the sensor or the recorder outputs are fed to the modulator (different modulations are in use, for example binary phase shift keying (BPSK) or quadrature phase shift keying (QPSK)) and, after amplification, are passed to the antenna. With conventional antenna designs, downlink rates of about 50 Mbps are in wide use. But there are already more sophisticated systems being used and studies in progress that currently use, or strive for, higher data rates. For instance, the spacecraft antenna may be designed as a high-gain antenna with a narrow radio frequency beam. In this case the spacecraft must be slewed to track the ground station during transit. This design is suitable for the store and dump option (option 2 above), or, if imaging takes place in the vicinity of the ground station, also for the real-time downlink option (option 1). This configuration does not support option 3, i.e. simultaneous imaging and downlinking in different directions within the ground station coverage circle. In this case only store and dump later is possible (option 2).

The solution to this problem is an autonomously steerable high-gain antenna: studies show that this system provides a greater than tenfold improvement in the data link and can be used even on board small satellites [da Silva Curiel, 2008]. HRS on SPOT-5 with a swath width of 60 km and GSD = 5 m has a sensor data rate of 70 Mbps. SPOT-5 with its 2 × 75 Mbps downlink rate can transmit the images from its two HRSs simultaneously. Also WorldView-1 can image and downlink with 800 Mbps a maximum contiguous area of 60 km × 110 km (mono) or 2 × 30 km × 110 km (stereo), in a ground station contact time of close to 7 minutes.

3.6.4 Possible ways of reducing data volume and downlink rates

3.6.4.1 Data compression

The first method to reduce data volume and the related downlink rate is to use data compression. The basic idea underlying data compression is the reduction of the volume of data by eliminating redundancy or less-relevant portions of the data. The algorithms used for this purpose can be divided into two main categories: the lossless and the lossy methods. Lossless methods convert a data set into a form in which there is less redundancy and a higher degree of compactness and which requires less storage capacity. Through appropriate decompression, the original data set can be completely restored. The compression factor in the case of lossless methods for spaceborne mapping systems is approximately two.

Lossy methods are used when part of the original data set can be eliminated without this being noticeable in the restored data set or having a detrimental effect on it. Hence, in the case of lossy data compression, the decompressed data are no longer identical to the original data, but they are still similar enough so that the observer does not notice any detrimental changes. The degree to which decompressed data may differ from the original depends to a large extent on the application in question; this has to be decided on a case-by-case basis. In the case of lossy methods, the compression factor may be as high as five without creating significant artefacts or image distortions.

Lossless methods

Lossless data compression removes redundancy from the data and codes them in a more compact form, without loss of information. The following two methods representative of this family are commonly used.

Run-length encoding: In the case of run-length encoding, like symbols such as bytes are counted and replaced by a single symbol and the corresponding numerical value. The numerical value indicates the number of times the symbol needs to be used to restore the original symbol series. Special characters in the data stream mark the places where such a symbol/number pair was used. This method is especially useful for black and white drawings with homogeneous areas, but it is not so well suited to photogrammetric applications, because they contain several different signal values and noise.

Huffman coding: Lossless methods using statistical distribution of symbols such as bytes for optimal coding of a data set are better suited for photogrammetric applications. Symbols that occur frequently in a data set are replaced by a shorter code word, whereas symbols that occur more seldom are replaced by a longer code word. Hence, encoding in this context means substituting a symbol by another, optimised bit sequence. As a result, less frequent symbols can be represented by longer bit sequences than original ones. But the data set as whole is represented in a more compact manner, because frequently occurring symbols are represented in an abbreviated form. Since the encoding key is needed for decompression, it is stored along with the encoded data. Although additional storage space is needed for this purpose, it is still smaller than the space saved by encoding. A code tree that can be implemented easily and efficiently is used for decompression.

Lossy methods

Lossy methods are based on the assumption that a data set contains information that can be eliminated without this later being noticed by the observer or having a distorting effect. By virtue of its standardisation, JPEG (Joint Photographic Experts Group) was for many years the most common method of compressing image data. There are other methods that have been developed and partly in use, for instance JPEG2000, which is based on the discrete wavelet transform and can be used for lossy and lossless compression. But JPEG may serve as an example to describe the principal approach.

JPEG works with image blocks of 8 × 8 pixels. These blocks are individually transformed into a frequency range, with the aid of a discrete cosine transformation (DCT), and the calculated 8 × 8 frequency portions are quantified by dividing each of the 8 × 8 portions by a certain value that is specified in an 8 × 8 quantisation table. Quantisation is achieved by dividing each of the 8 × 8 frequency ranges by a certain value specified in the 8 × 8 quantisation table. Higher-frequency portions are smoothed as a result of division with reduced precision. The quantified frequency portions are then processed separately as AC and DC values, compressed in a lossless manner by means of Huffmann coding, formatted and stored. The quantisation table can be initialised using one's own values or those suggested in the JPEG Standard. One should bear in mind that information content is removed in the quantisation process, because the precision of the quotient that is stored is diminished. The value of the quantisation coefficient determines the amount of data that is lost, so the quality loss of an image can be controlled in this manner. For lossy methods, just as in the case of lossless methods, certain data have to be stored for the decompression, in addition to image data. In the decompression process, the data set undergoes the same stages of compression, but in the reverse order. However, because of quantisation, the decompressed data set is no longer identical to the original data after decompression.

The maximum compression rate used in photogrammetric applications is five, which ensures that no geometric distortions occur when using the JPEG method. Using JPEG, we map the sensor data from the spatial domain into the frequency domain. This way we can easily understand that we can get much better compression results if we can get rid of the high-frequency data artefacts originating from the photo response non-uniformities (PRNU) of the adjacent detector elements. The PRNU correction is described in Section 3.6.4.2. Looking at the compression from the data rate and volume reduction aspect only, it is best to do it online before the storage device.

3.6.4.2 Extracting the information content from the sensor data

The second method to reduce the sensor raw data volume is to extract the information content from the sensor data. With reference to all components of the spaceborne camera, the signal S_i at the optoelectronic converter element i is described by

$$S_i = I_K \int_{t_{int}} \int_{\Delta\lambda} \left[R_{CCD}(\lambda) \, . \, T_{Opt}(\lambda) \, . \, L_i(\lambda) d\lambda + DS \right] dt \tag{3.16}$$

where I_K is a camera-dependent constant (focal length << flight altitude)

$$I_K = \frac{\pi}{4} \frac{A_d \, . \cos^x (\vartheta_i)}{K \, . (\frac{f}{D})^2} \tag{3.17}$$

and

t_{int} = integration time,
$\Delta\lambda$ = current wavelength range,
R_{CCD} = CCD spectral responsivity,
T_{Opt} = spectral transmittance of lens and filter,
L_i = spectral radiation flow in front of the lens,
DS = dark signal generation rate,
x = lens-dependent exponent of the "$\cos^4$ law", $2 < x < 4$,
θ_i = visor angle of the pixel i,
f/D = F-number,
K = conversion factor of the CCD,
A_d = detector area,
f = focal length.

In practical work, fixed integration times and values averaged in the observed wavelength range are used, i.e.

$$R_{CCD}(\lambda) \, . \, T_{Opt}(\lambda) \, . \, L(\lambda) \approx R'_{CCD} \, . \, T'_{Opt} \, . \, L' \tag{3.18}$$

from which results the useful equation

$$S_i = I_K (R'_{CCD} \, . \, T'_{Opt} \, . \, L' + DS) t_{int} \tag{3.19}$$

Equation (3.19) is well suited to describing how the extraction of information from the sensor raw data works. The signal S_i at the detector element i is a function of the radiation quantity L' in front of the lens and of a dark signal DS, dependent on the detector.

The temperature-dependent dark signal *DS* consists of the mean value and DSNU (dark signal non-uniformity) value and has to be subtracted; this is a systemic error term that is dependent on the structural components. Hence, it is determined during camera calibration and stored in the correction value memory.

With every spaceborne camera, the aim is to achieve the best possible signal-to-noise ratio (SNR) in order to be able (at all processing stages) to work with data that are as close as possible to the information they contain. There are many components in the signal chain that affect noise in the digital spaceborne camera. If we restrict ourselves to producing a rough breakdown of the noise components, the SNR can be described by

$$SNR = \frac{n_s}{\sqrt{\sigma_s^2 + \sigma_{fp}^2 + \sigma_{rms}^2}} \tag{3.20}$$

where n_s is the number of electrons collected in the CCD, generated by the signal S_i; σ_s^2 is the variance of the signal electrons *ns*; σ_{fp}^2 is the variance of the local responsivity differences (fixed pattern noise); and σ_{rms}^2 is the variance of the time-dependent noise of all other components involved.

The number of signal electrons is proportional to the number of photons impinging on the detector in the spectral band in question. Hence, the noise of the signal electrons also follows the Poisson distribution, i.e.

$$\sigma_s = \sqrt{n_s} \tag{3.21}$$

For instance, the time-dependent noise of the CCD and of the analogue data channel (rms noise) contains

- the dark signal noise (Poisson distribution);
- the noise of the reset and amplifier circuits ("kTC noise");
- the charge transfer noise;
- other noise components (1/*f* noise, thermal noise);
- noise response of the A–D converter.

The causes of the fixed pattern noise σ_{fp} are the responsivity differences (PRNU) of the individual detector elements, and the drop in light intensity towards the edge of the focal plane of a (wide-angle) lens (so-called "$\cos^4$ law"). Figure 3.20 shows the typical curve of the output signal of a CCD line (individual line or line of a matrix) with approximately 5200 elements, whose centre element is situated on the focal plane in the vicinity of the optical axis of a wide-angle lens. The curve in Figure 3.20 represents the two portions of fixed pattern noise σ_{fp}: the $\cos^4$ drop in light intensity towards the edges of the focal plane (shading) behind a wide-angle lens, and the photo response non-uniformity of the CCD elements. Of course, for narrow-angle optics, this drop is much smaller but canot always be neglected. Both portions are actually systemic errors, but they can be regarded as noise portions if they are not recognised and corrected.

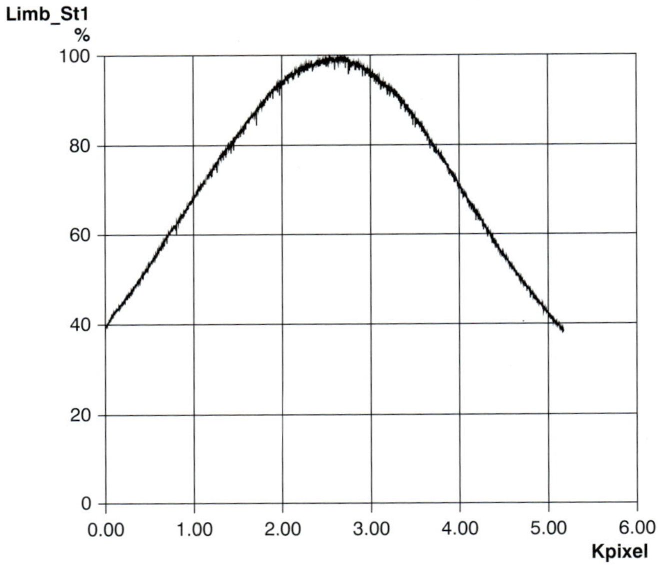

Figure 3.20 Typical curve of an output signal of a CCD line in the focal plane of a wide-angle lens, given a homogeneous intensity distribution of illumination.

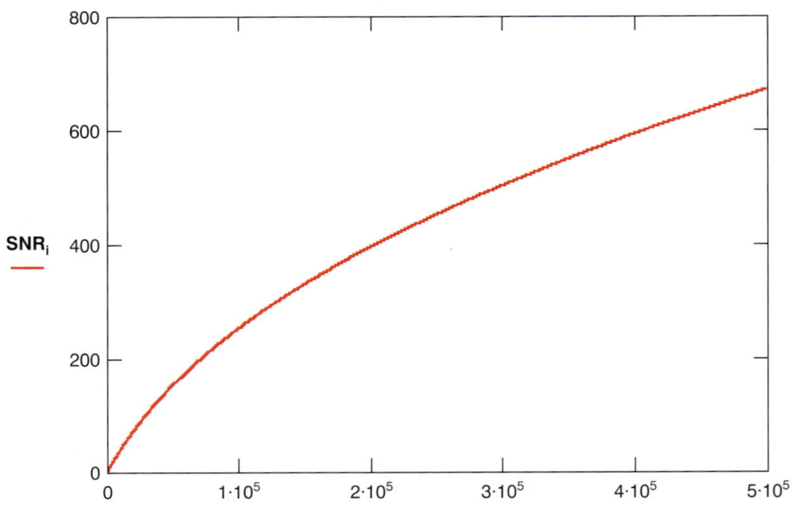

Figure 3.21 Modulation-dependent SNR of a CCD element with a saturation capacity of 500 000 e$^-$ and a σ_{rms} of 200 e$^-$.

If a selected CCD element is considered (e.g. an element near the optical axis) and σ_{fp} is ignored for this selected element, the SNR curve shown in Figure 3.21 is obtained for this CCD element, given a saturation capacity of 500 000 e$^-$ (electrons) and a σ_{rms} of 200 e$^-$. For saturation charges of 100 000 e$^-$ to 500 000 e$^-$, we obtain an SNR ranging from 267 to 680, which corresponds to 8 and 9 bits, respectively, regardless of how many bits are used or necessary for image collection. It is important to note that, even in an ideal noiseless system, because of the Poisson statistics of the incoming photons, the SNR cannot exceed the square root of the saturation capacity of a detector element.

The question is, when does the effect of PRNU have to be corrected, and how precise must this correction be? In CCD data sheets, PRNU is usually indicated as a percentage of the output signal below saturation in the linear part of the characteristic curve. The fixed pattern noise of the responsivities of the CCD elements generated by PRNU, which changes into a time-dependent noise as a result of charge transfer, can be expressed by

$$
\begin{aligned}
\sigma_{fp} &= \frac{PRNU}{100\%} \cdot n_s \\
&= \frac{PRNU}{100\%} \cdot \sigma_s^2
\end{aligned}
\tag{3.22}
$$

As a function of the concrete PRNU of the CCD signals, which also includes the shading part of a narrow-angle optics here, the SNR (from Equation 3.20) is

$$
SNR = \frac{\sigma_s}{\sqrt{1 + \left[\frac{PRNU}{100\%}\sigma_s\right]^2 + \left[\frac{\sigma_{rms}}{\sigma_s}\right]^2}} \tag{3.23}
$$

Figure 3.22 shows the best possible SNR, given full modulation, again assuming a saturation charge of 500 000 e$^-$ and σ_{rms} = 200 e$^-$.

Up to a PRNU of 0.02%, the SNR is determined by the photon noise of the input radiation and σ_{rms}. For PRNU above 0.1%, the influence of PRNU then dominates. Since the PRNU values of CCD components are currently in the region of a few per cent, a pixel-by-pixel PRNU (including shading) correction is required. The signals corrected in this manner ensure good SNR values over the entire dynamic range *DR* of the camera

$$
DR = \frac{S_{max} - DS \,.\, t_{int}}{\sigma_{rsm}} \tag{3.24}
$$

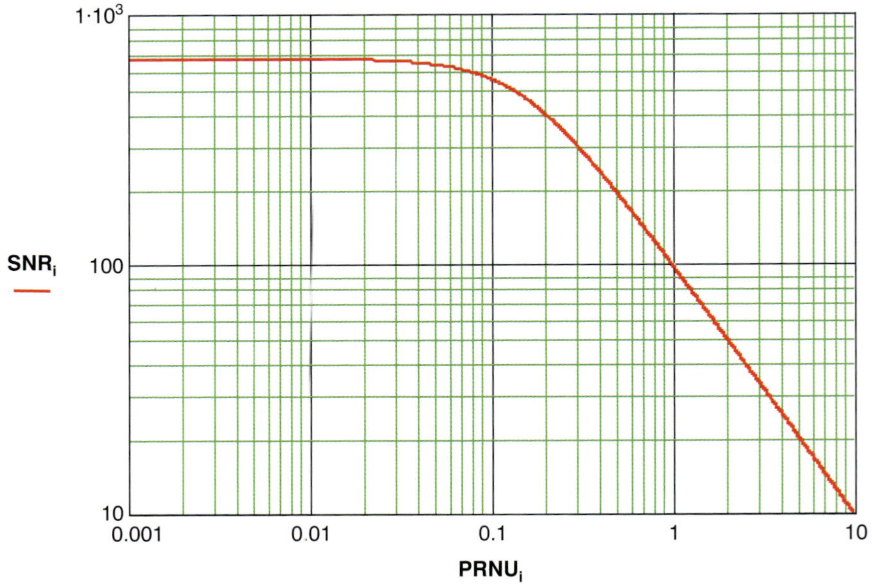

Figure 3.22 SNR as a function of PRNU for a saturation charge = 500 000 e^- and σ_{rms} = 200 e^-.

Since the temperature-dependent DS can easily be subtracted, the term $DS \times t_{int}$ can be neglected. Furthermore, applying the data compression algorithm to these information data gives much better results since the high-frequency part of PRNU has no influence on compression performance.

References

Boissin, B., Baudoin, A., Begni, G., Fontannaz, D. and Munier, P., 2002. A new generation satellite: SPOT-5 in orbit. *MEDIAS Newsletter*, (13), 74–77.

Brieß, K., Jahn, H., Lorenz, E., Oertel, D., Skrbek, W. and Zhukov, B., 2003. Fire recognition potential of the Bi-spectral InfraRed Detection (BIRD) Satellite. *International Journal of Remote Sensing*, 24(4), 865–872.

Cutter, M., Lobb, D., Purll, D., Cawthorne, A. and Sweeting, Sir Martin, 2008. The trade-off of high resolution imagers on small satellites. *Proceedings 59th International Astronautical Congress*, Glasgow, Scotland.

DigitalGlobe, 2009. DigitalGlobe Constellation – WorldView-1 Imaging Satellite, DS-WV1Rev12/09.

Jacobsen, K., 2005. High resolution satellite imaging systems an overview. *Photogrammetrie, Fernerkundung, Geoinformation*, 6, 487–496.

Jahn, H. and Reulke, R., 1995. *Systemtheoretische Grundlagen optoelektronischer Sensoren*, Akademie Verlag, Berlin, Germany.

Konecny, G., 2003. *Geoinformation – Remote Sensing, Photogrammetry and Geographic Information Systems*. Taylor & Francis, London and New York.

Kramer, H.J., 2002. *Observation of the Earth and its Environment: Survey of Missions and Sensors*, 4th edn. Springer Verlag, Berlin, Heidelberg and New York.

Krause, K., 2005. Radiometric use of QuickBird imagery, DigitalGlobe Technical Note, http://www.digitalglobe.com/downloads/products/QuickBird_technote_raduse_v1.pdf (accessed March 2011).

Morris, N., Edeson, R., Waltham, N., Tosh, I., Tyc, G. and Butlin, T., 2008. One metre resolution remote sensing camera for small satellites. *Proceedings 59th International Astronautical Congress*, Glasgow, Scotland.

Sandau, R., 2004. Metrics to support the design of small-satellite-based high-resolution mapping systems. 18th Annual AIAA/USU Conference on Small Satellites, August 9–12, Logan, UT, USA.

Sandau, R. (Ed.), 2010. *Digital Airborne Camera: Introduction and Technology*. Springer, Dordrecht, Heidelberg, London and New York (original German edition published by Wichmann Verlag, 2005).

da Silva Curiel, A., 2008. Increasing the data volume returned from small satellites. *Proceedings 59th International Astronautical Congress*, Glasgow, Scotland.

Spilker, J.J., 1997. *Digital Communications*. Prentice-Hall Inc., Englewood Cliffs, NJ, USA.

4 Sensors with a GSD of greater than 1 m up to 16 m

Karsten Jacobsen and Ian Dowman

4.1 Introduction

This chapter describes the sensors launched from 1986 onwards, with a pixel size of greater than 1 m and less than 16 m, starting with SPOT-1. Most of these have the characteristic of collecting stereoscopic data either by pointing sideways and collecting data from more than one orbit, or by having two or three sensors with some combination of forward-, nadir- or backward-pointing telescopes; a few of them are agile with stereo mapping possible by rotating the satellite. All use push-broom scanning technology with the exception of the experimental UoSAT-12, which is equipped with CCD matrices.

The chapter starts with some general information on the effects of oblique pointing and then gives details of individual sensors and the results obtained.

4.2 General information

SPOT-1 collected stereoscopic data from two orbits by sideways pointing, as shown in Figure 4.1(a). Later systems, such as MOMS, ASTER, Cartosat-1 and ALOS PRISM collected data by forward-, nadir- or backward-pointing as shown in Figure 4.2(b). Both of these systems required oblique viewing.

The stereoscopic coverage from different orbits (Figure 4.1a) has the disadvantage of a non-negligible time interval between both images, which may affect the radiometric similarity of the images. In addition, the scenes do not cover the same area of ground because of the convergent orbits. All listed optical satellites have sun-synchronous orbits, imaging between 10:00 and 11:00 o'clock in the morning. Images cannot be taken from another orbit until at least a day later. Often the configuration is not optimal on the next day and very often the area is covered by clouds, causing a time interval between both images of several days or even a month. During this time, especially in agricultural areas, the object may change and the sun elevation is also changing, leading

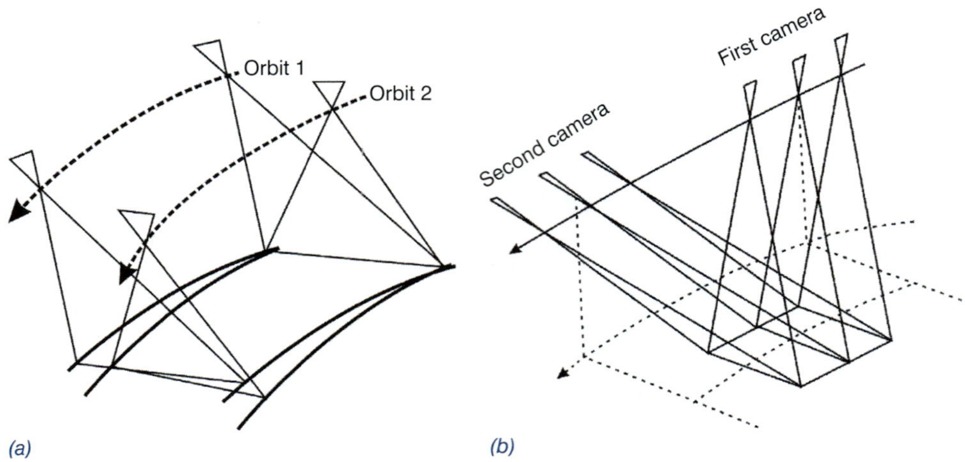

Figure 4.1 Modes of stereoscopic viewing: (a) stereoscopic imaging from different orbits; and (b) imaging from the same orbit with stereoscopic systems.

to different shadows. The automatic image matching of such a stereo pair for generating height models may be difficult or impossible. Such problems do not exist for the stereo camera systems or agile satellites, which image the same area approximately one minute later (Figure 4.1b).

Sensors are classified by their ground sampling distance (GSD). Neighbouring pixels may be under-sampled or over-sampled if the sensor is not pointing to the nadir, which means that the sensor element projected onto the object space will not be the same as the nominal GSD. If the sensor has only a single CCD line, in the across-track direction with a nadir-pointing array, the projected pixel size is identical to the GSD for original images (basic imagery). In the along-track direction, the GSD is determined by the sampling rate – the number of lines per second – and the footprint speed of the satellite. For a view not in the nadir direction, the projected pixel size depends on the incidence angle i (Figure 4.2a) – the nadir angle from the ground to the satellite, which is larger than the nadir angle at the satellite caused by the Earth's curvature (Figure 4.2a).

Figure 4.3 shows how pixel size varies as a function of the incidence angle according to the formulae:

$$P_v = P/\cos^2 i \tag{4.1}$$

$$P_o = P/\cos i \tag{4.2}$$

where P_v is the projected pixel size in the view direction; P_o is the projected pixel size in the across-track view; P is the pixel size in nadir view; and i is the incidence angle.

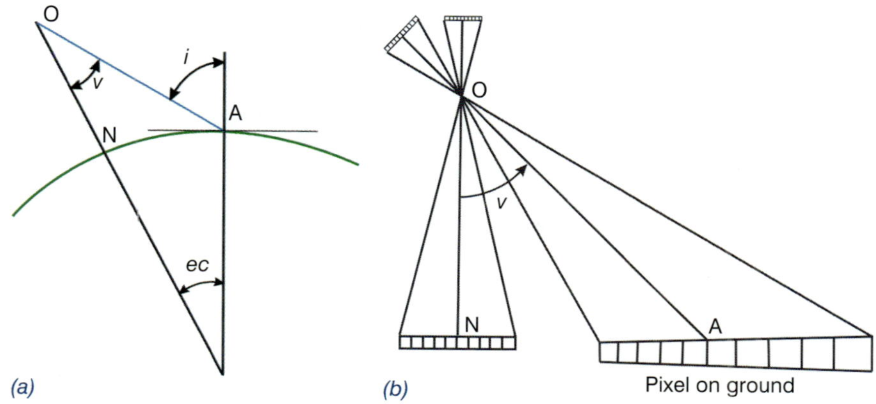

Figure 4.2 The effect of incidence angle on pixel size: (a) nadir angle v, incidence angle i, Earth centre angle ec; and (b) sensor elements projected onto the ground.

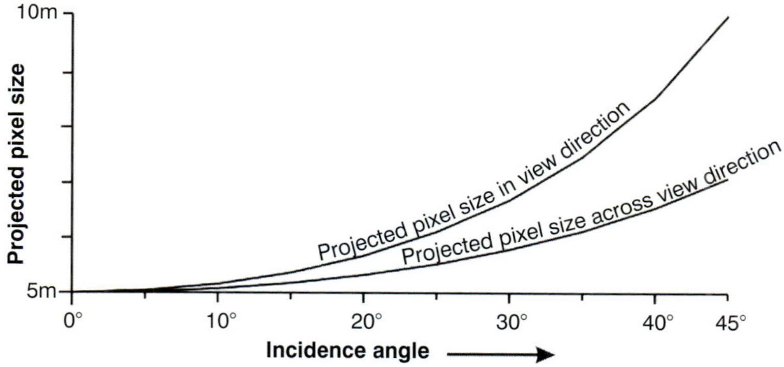

Figure 4.3 Projected pixel size as a function of incidence angle for 5 m GSD in nadir view.

For optical sensors like the HRV on SPOT 1–4 and HRG on SPOT-5, viewing only across the orbit, the projected size in the view direction determines the GSD across the orbit, while in the orbit direction the GSD is determined by the sampling rate of 5.0 m for the panchromatic image from SPOT-5. For satellites with flexible view direction, the projected pixel size in orbit direction causes an over-sampling, reducing the image quality.

Ground resolution can be improved by using two staggered CCD lines instead of single CCD lines (Figure 4.4a). Staggered CCD lines are shifted half a pixel relative to each other. In this case the GSD is linearly related and half the size of the projected pixel size (Figure 4.4b). See also Section 3.5.3.

The 50% over-sampling influences the image quality slightly, but the quality or the information contained therein (elements which can be identified in the image) is still better than with a GSD twice as large. A typical example of a sensor with staggered

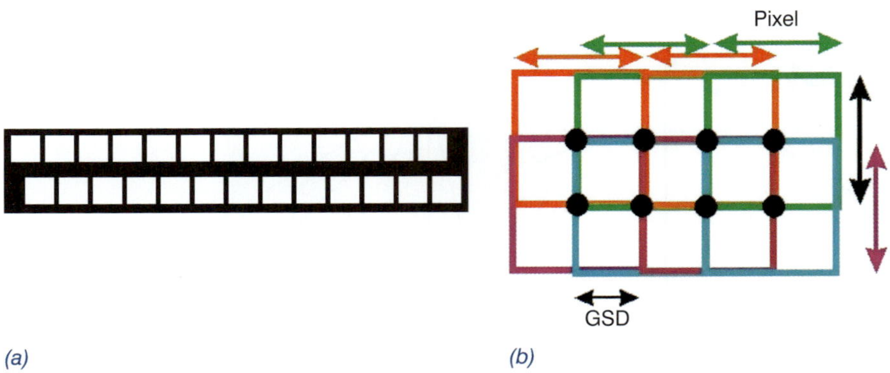

Figure 4.4 (a) Staggered CCD lines; and (b) the relationship of GSD to the projected pixel size for staggered CCD lines.

CCD lines is ASTER, having 15 m GSD, but 30 m projected pixel size, and SPOT-5 supermode. SPOT-5 supermode is using even less information as shown in Figure 4.4(b) (see also Figure 4.12) for reaching 2.5 m GSD with 5 m pixel size. The main reason for using the larger pixels of staggered CCD lines, rather than pixels with half the size in both directions, is the increase in the light sensitivity and the reduction of the signal-to-noise ratio (SNR).

Another possibility for collecting more light for each sensor element is the asynchronous imaging mode. In this mode the view direction is changed continuously, enlarging the exposure time for each ground element by the factor *B*/*A* (Figure 4.5). For example, the sampling rate of EROS A is limited to 750 lines per second. With a footprint speed of 7088 m s^{-1}, a slow-down factor of 5.2 is required for 1.8 m GSD. Combinations of different methods are also possible.

With a footprint speed of approximately 7000 m s^{-1}, only 0.28 ms are available for the exposure of one pixel having 2 m projected pixel size. A limited aperture of the optical systems means this is not a sufficient exposure time under usual lighting conditions and for an acceptable SNR. There are three possibilities for increasing the combined energy to the sensor element: 1) an asynchronous imaging mode with permanent change of the attitude against the orbit; 2) staggered CCD lines; and 3) transfer delay and integration (TDI) (Figure 4.6).

A TDI sensor has a small CCD array. With the speed of the motion of the projected object in the image plane, the energy collected in the first CCD line is shifted in the form of free electrons to the neighboured CCD line, where more energy is collected. The summed free electrons are shifted to the next CCD line and so on. IKONOS and QuickBird are equipped with CCD arrays of 64 lines, so in theory the exposure time can be extended 64 times. This effect does not influence the imaging geometry – this corresponds to the geometry of the centre CCD line. But, in the case of a lateral tilt of

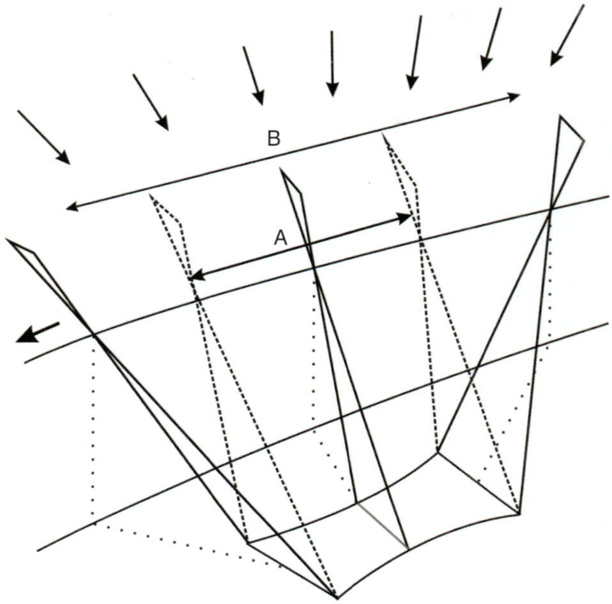

Figure 4.5 Asynchronous imaging mode with slow-down factor B/A. Dotted lines with base A corresponds to view direction not changed during imaging.

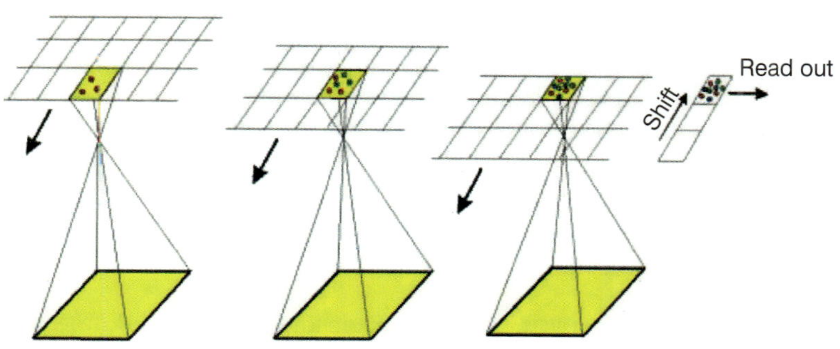

Figure 4.6 Time delay and integration (TDI): image motion and synchronous shift of generated free electrons in the CCD array.

the view direction, the projected object has an image motion component also in the line direction, causing a reduced radiometric image quality. For this reason, TDI is usually limited to 13 steps in IKONOS and QuickBird.

The radiometric resolution can be investigated by edge analysis. A sudden change of the brightness in the object from one location to a neighbouring area (Figure 4.7a, top), for example from a bright roof to a dark shadow, is causing a continuous change of the grey value profile in the image (Figure 4.7a, bottom). The grey value profile

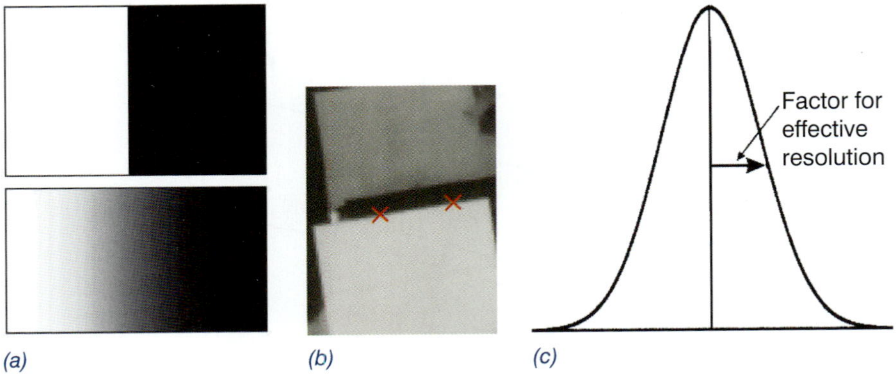

Figure 4.7 The use of edge analysis to determine the effective resolution: (a) edge in object space (top) and edge in image space (bottom); (b) analysed edge; and (c) point spread function.

can be differentiated, leading to the point spread function (PSF) in Figure 4.7c. The width of the PSF gives information on the resolution; with digital images this will be normalised to the GSD. Half of the width of the PSF will be taken as the factor for effective resolution. This factor multiplied by the nominal GSD leads to the effective GSD corresponding to the real information content of the image. If this factor for the effective resolution is equal to 1.0, the effective image resolution corresponds to the nominal GSD. In the case of SPOT supermode images, the factor for the effective resolution is approximately 1.2, so the effective resolution is the GSD multiplied by this factor, corresponding to 2.5 m . 1.2 = 3.0 m, so the image includes information corresponding to images with 3.0 m GSD. In theory this factor should not be below 1.0, but a factor below 1.0 can be caused by contrast enhancement.

Usually only images corrected by the geometric and radiometric camera calibration are distributed. The corrected scenes are named "level 1A" by Spot Image or "basic images". In addition, very often images projected onto a plane of constant height are used, which may be related to the national or international reference system: these are named "level 2A" by Spot Image or "IKONOS Geo" or "QuickBird OR standard" (Figure 4.8). See also Section 1.2. Level 2A type images are usually georeferenced, but they are not orthoimages related to the actual terrain height, and the orientation is based on the direct sensor orientation of the satellite, which has to be improved by means of control points. A height difference against the plane of rectification causes a horizontal shift by:

$$\Delta L = \Delta h \,.\, \tan i \tag{4.3}$$

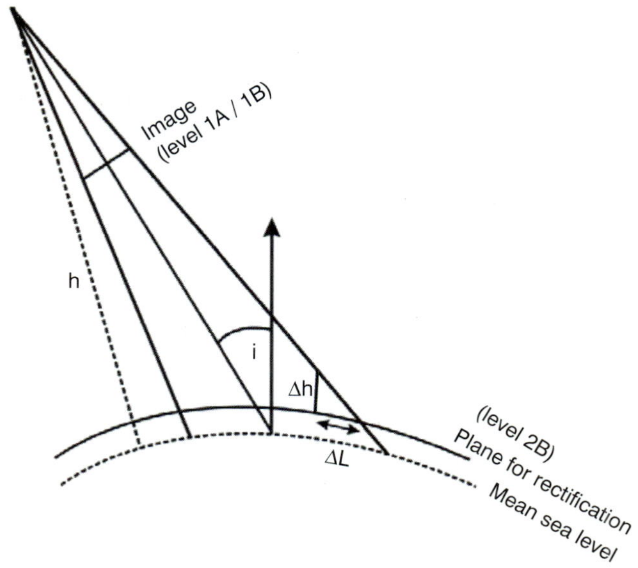

Figure 4.8 Distributed image types: level 1A type is close to original image; level 2A type is rectified to a plane of constant height.

4.3 Sensors

4.3.1 Overview

Table 4.1 shows the sensors in this category.

In addition to the listed digital satellites, old analogue systems, formerly used by the military but now open for civilian use, have to be mentioned. Images taken by CORONA can now be ordered via the Internet for a handling fee; these images are from the 1960s, but are useful for change detection and archaeological purposes. This is also the case for the Russian space photographs which have been taken up to approximately the year 2000. The Russian KFA-1000, KFA-3000 and KVR-1000 belong to the resolution group under consideration in this chapter. Section 4.4 deals with photographic systems.

The use of optical satellite images for civilian mapping started with the launch of SPOT-1 in 1986. Military reconnaissance used digital optical satellites earlier than this, but they are only mentioned here if the images taken are accessible today for civilian applications. The 10 m panchromatic GSDs of SPOT-1 up to SPOT-4 were only useful for small-scale mapping; this was not real competition for aerial images. With improved ground resolution this has changed; space images achieving 5 m GSD are able to compete with aerial images, and the decision on whether to use aerial or space images for topographic mapping is now based on financial constraints and image availability.

Table 4.1 Details of sensors with pixel size of greater than 1 m and less than 16 m

Sensor	Agency Country launch date	Platform altitude (km)	GSD pan/MS (m)	Pointing capability (°)	Swath width at nadir (km)	Channels (µm)	Quantisation (bits)
HRV High-resolution visible	Spot Image France 1986, 1990, 1993, 1998	SPOT 1–4 832	10/20	sideways up to ±27	117 (i.e. 60 + 60 with 3 overlap)	pan 0.51–0.73 MS: 0.5–0.59, 0.61–0.68, 0.79– 0.89	11
MOMS-02 Stereoscopy	DLR Germany 1993	Space Shuttle 296	4.5/13.5	forward/ nadir/ aft −21.4/0/21.4	37/78	pan 0.52–0.76 MS: 0.44–0.505, 0.53–0.575, 0.645–0.68, 0.77–0.81	7
MOMS-2P Stereoscopy	DLR Germany 1996	MIR space station 400	6/18	forward/ nadir/ aft −21.4/0/21.4	48/101	spectral range as MOMS-02	7
PAN (IRS-1C/1D) Panchromatic sensor	ISRO India 1995/1997	IRS-1C/IRS-1D 817/803	5.8	sideways up to ±26	70	pan 0.5–0.75	6
ASTER Advanced Space-borneThermal Emission and Reflection Radiometer Stereoscopy	NASA, METI USA, Japan 1999	Terra 705	VNIR: 15 stereo: 15 horizontally and 25 vertical SWIR: 30 TIR: 90	Nadir / aft 0°/27°	60	stereo in NIR MS: 3 bands in 0.52–0.86 SWIR: 6 bands in 1.6–2.43 TIR: 5 bands in 8.125–11.65	8

UoSAT12	UK 1999	650	10/32	free view	10 m pan 32 m MS	CCD matrix 1024 × 1024 pixels	8
EROS A1	ImageSat Int. Cyprus (Israel) 2000	480	1.8	free view	12.6	pan 0.5–0.9	11
HRS High-resolution stereoscopy	Spot Image France 2002	SPOT 5 830	5 × 10	forward/ aft +20/–20	120	pan 0.49–0.69	8
HRG [a]	Spot Image France 2002	SPOT 5 830	5(2.5)/10	sideways up to ±27	60	pan 0.49–0.69 MS: 0.50–0.59, 0.61–0.68, 0.79–0.89, 1.50–1.75	8
LISS-IV Linear imaging self-scanner-IV	ISRO India 2003, 2008	Resourcesat-1 Resour-cesat-2 817	5.8	sideways up to ±26	70	MS: 0.52–0.59, 0.62–0.68, 0.77–0.86	10
FORMOSAT 2	NSPO China, Taiwan 2004	890	2/8	free view	24	pan 0.45–0.90 MS: 0.45–0.52, 0.52–0.60, 0.63–0.69, 0.76–0.90	12
PAN (Cartosat-1) Panchromatic sensor Stereoscopy	ISRO India 2005	Cartosat-1 613	2.5	forward/ aft 26/5 free view to side up to 23	27	pan 0.5–0.75	10

(Continued)

Table 4.1 (cont.) Details of sensors with pixel size of greater than 1 m and less than 16 m

Sensor	Agency Country Launch	Platform altitude (km)	GSD pan/MS (m)	Pointing capability (°)	Swath width at nadir (km)	Channels (μm)	Quantisation (bits)
TopSat Telescope	BNSC UK 2005	TopSat 686	2.8/5.6	free view	15 m pan 10 m MS	pan 0.50–0.70 MS: 0.4–0.5, 0.5–0.6, 0.6–0.7	8
PRISM Panchromatic Remote-sensing Instrument for Stereooscopy	JAXA Japan 2006	ALOS 699	2.5	forward/ nadir/aft –24/0/+24 free view to side	70 35 (triplet stereo observations)	pan 0.52–0.77	8
AVNIR-2	JAXA Japan 2006	ALOS 699	10	free view to side up to 45	70	0.42–0.50, 0.52–0.60 0.61–0.69, 0.76–0.89	8
PAN (BJ-1) Panchromatic Imager	NRSCC (CAST) 2005	BJ-1 Beijing 1 686	4/32	free view	24/640	pan 0.50–0.80	8
CBERS-2B	CNSA / INPE China / Brazil 2007	778	2.4/20	free view	27 Pan 113 MS	pan 0.51–0.73 MS: 0.45–0.52, 0.52–0.59, 0.63–0.69, 0.77–0.89	8

MSI Multi-spectral imager	RapidEye Germany 2008	RapidEye 5 satellites 530	6.5	free view	78	MS: 0.44–0.51, 0.52–0.59, 0.63–0.685, 0.69–0.73, 0.76–0.85	12
THEOS	GISTDA Thailand 2008	822	2/15	free view	22 Pan 90 MS	pan 0.45–0.90 MS: 0.45–0.52, 0.53–0.60, 0.62–0.69, 0.77–0.90	8 from 12
AISat-2	Algeria 2008	680	2.5/10	up to 30 cross track	17.5	pan 0.45–0.90 MS: 0.45–0.52, 0.53–0.60, 0.62–0.69, 0.77–0.89	10

[a] Two HRG instruments on this mission can be processed to produce simulated imagery of 2.5 m.

4.3.2 SPOT range of sensors

4.3.2.1 HRV on SPOT 1–4

The launch of SPOT-1 in February 1986, with two high resolution visible (HRV) sensors, was the beginning of civilian mapping using optical space images. The resolution and configuration of the earlier Landsat sensor was designed for land classification – not for generating 3D data or appropriate for object positioning. With 10 m GSD for the panchromatic channel, the ground resolution of SPOT-1 was still limited, not really competing with aerial images, but opened its application in areas with difficult access. Full technical details of the SPOT satellites can be found on the website [SPOT5, 2011].

SPOT 1–4 have two HRV sensors, which can point up to ±27° to the sides, as shown in Figure 4.9, thus allowing stereoscopic image pairs to be collected from different orbits (Figure 4.10). The telescope of the HRV sensor for the panchromatic band has 12 000 CCD elements, each 6.5 µm square, and has a focal length of 1082 mm. The SPOT satellites use the DORIS positioning system (ftp://ftp.ids-doris.org/pub/ids/satellites/DORISSatelliteModels.pdf), which has been upgraded over the period of SPOT 1–5 as shown in Table 4.2 ("rmse" is the root mean square error). The pointing precision of SPOT-5 has also improved.

Table 4.3 shows the spectral bands of the SPOT sensors. Like most of the subsequent optical satellites, the ground resolution of the panchromatic channel is higher than for the multispectral channel. As with several other sensors, SPOT has no channel for the blue range: the short wavelength, strongly disturbed by moisture and dust in the

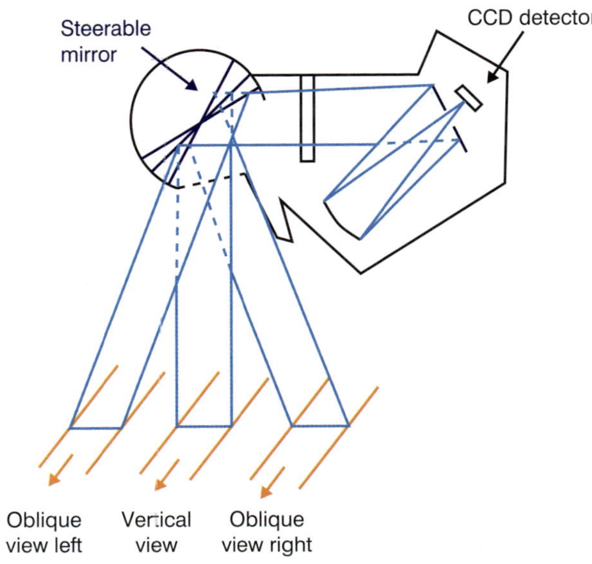

Figure 4.9 The SPOT HRV instrument.

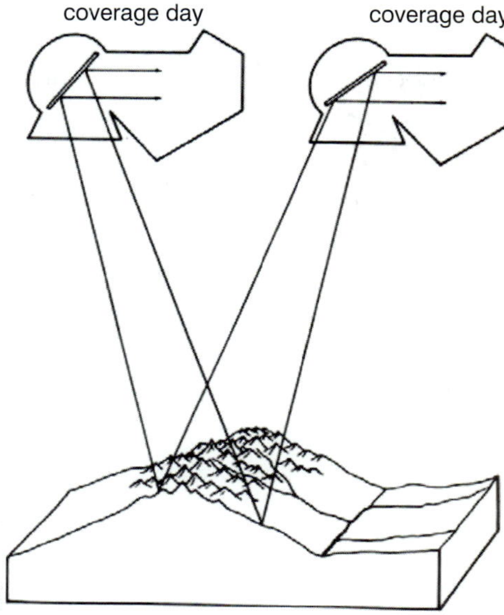

Figure 4.10 Stereo image collection with SPOT HRV. Credit: Spot Image

Table 4.2 Positioning systems on SPOT 1–5 (from http://spot5.cnes.fr/gb/satellite/satellite.htm)

	SPOT 1–3	SPOT-4	SPOT-5
Orbit determination	DORIS	DORIS in real time giving 5 m rmse	DORIS in real time giving 5 m rmse
Location accuracy (m)	350	350	50

Table 4.3 Spectral range of SPOT channels

	Pan (M) (μm)	Green (μm)	Red (μm)	NIR (μm)	SWIR (μm)
SPOT 1–3	0.50–0.73	0.50–0.59	0.61–0.68	0.78–0.89	–
SPOT-4	0.61–0.68	0.50–0.59	0.61–0.68	0.78–0.89	1.58–1.75
SPOT-5	0.48–0.71	0.50–0.59	0.61–0.68	0.78–0.89	1.58–1.75

atmosphere, includes only limited object information. On the other hand, since the introduction of SPOT-4, a short-wave infrared (SWIR) band has been included, improving the object classification. Instead of the panchromatic band, SPOT-4 has a red channel, called the "M band", with the high resolution of 10 m GSD. This was changed back to a panchromatic band for SPOT-5. The panchromatic band is almost identical to the visible range and is a disadvantage for mapping forest areas: in these areas the grey value distribution in the panchromatic band is very limited, making the automatic image matching for the generation of height models difficult.

4.3.2.2 HRG on SPOT-5

SPOT-5 has two high resolution geometric (HRG) instruments, which operate in a similar manner to the HRV in that they point sideways and have the same focal length and a virtual CCD line with 12 000 pixels (Figure 4.11).

The instruments have an additional CCD line shifted by half a pixel relative to each other, similar to standard staggered CCD lines (Figure 4.4). The sampling rate corresponds to 5 m GSD, so two images with the same number of pixels, shifted against each other by half a pixel in x and y directions, are produced. The so-called "supermode image" with 2.5 m GSD (Figure 4.12) can be generated from these independent images, by interleaving, interpolation and image restoration. The supermode images are not like original images with 2.5 m GSD, but include approximately the same information content as original images with 3 m GSD. So supermode images are clearly better than the original SPOT-5 images with 5 m GSD, which is also obvious by a simple visual comparison.

SPOT-5 generates the images without rotating the satellite against the orbit; only the mirror for viewing towards the ground is continuously changed to compensate for the influence of the Earth rotation on the imaged area – this is called "yaw control" (Figure 4.13).

4.3.2.3 HRS on SPOT-5

SPOT-5 also includes the high resolution stereo (HRS) autonomous stereoscopic sensor, enabling a stereoscopic coverage within the orbit, by viewing 20° ahead and 20°

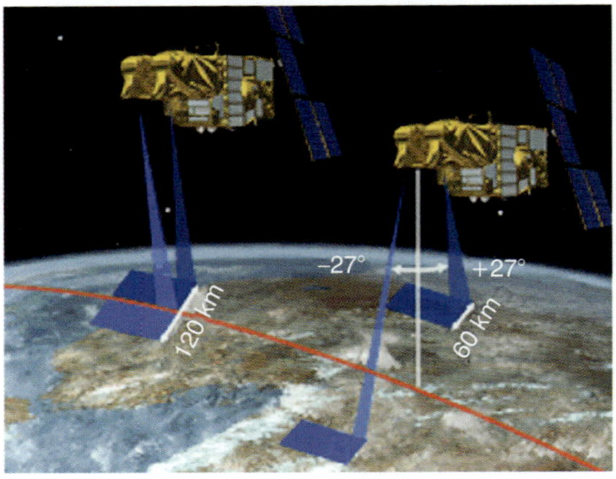

Figure 4.11 SPOT HRG operating modes. Credit: Spot Image

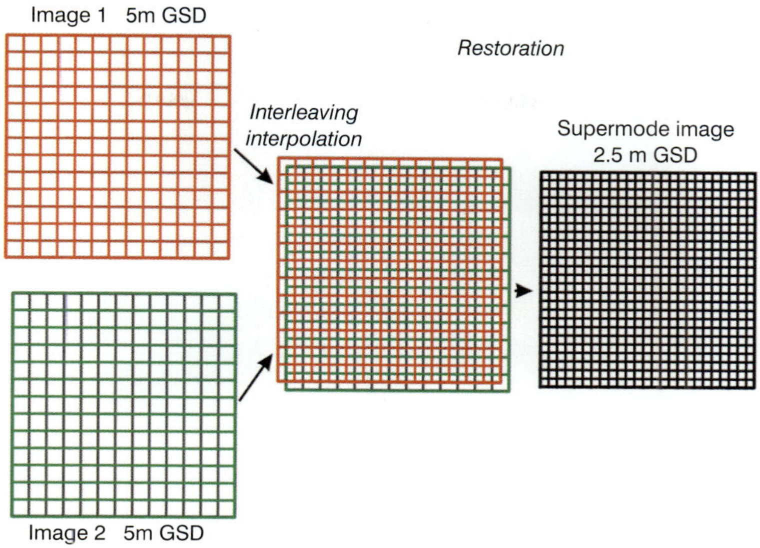

Figure 4.12 Generation of a SPOT-5 supermode image with 2.5 m GSD.

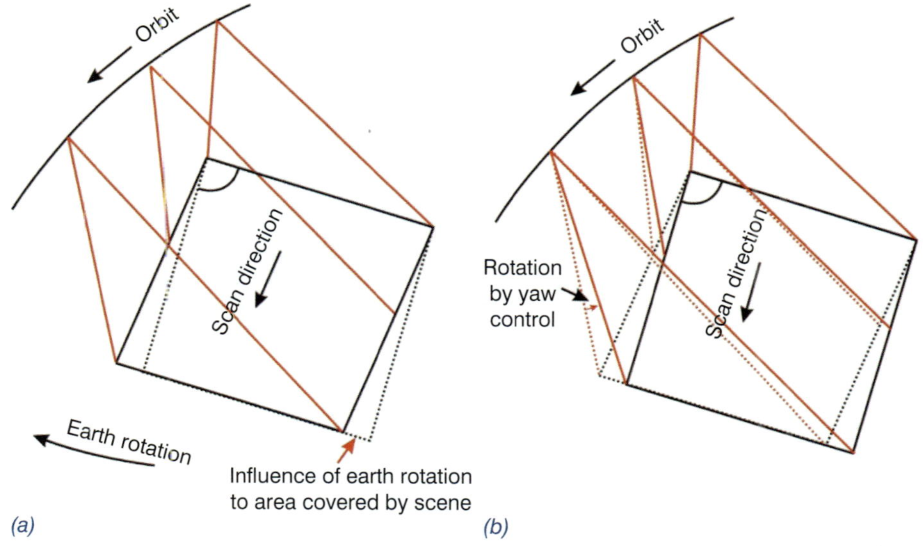

Figure 4.13 Yaw control on SPOT-5: (a) the area covered due to Earth rotation; and (b) SPOT-5 scene with yaw control, eliminating the influence of Earth rotation.

behind, corresponding to a ±22.5° incidence angle (Figure 4.14). Optimal conditions for height determination in open areas are given with a base-to-height ratio of 0.83.

The HRS camera system includes two combined CCD-line cameras with focal lengths of 0.58 m and arrays of 12 000 pixels covering a swath of 120 km, as shown in Figure 4.15. By an over-sampling in the flight direction a pixel size on the ground of

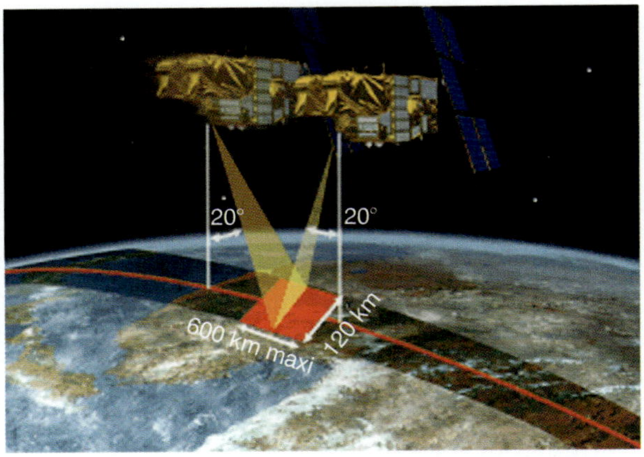

Figure 4.14 SPOT HRS operating mode. Credit: Spot Image

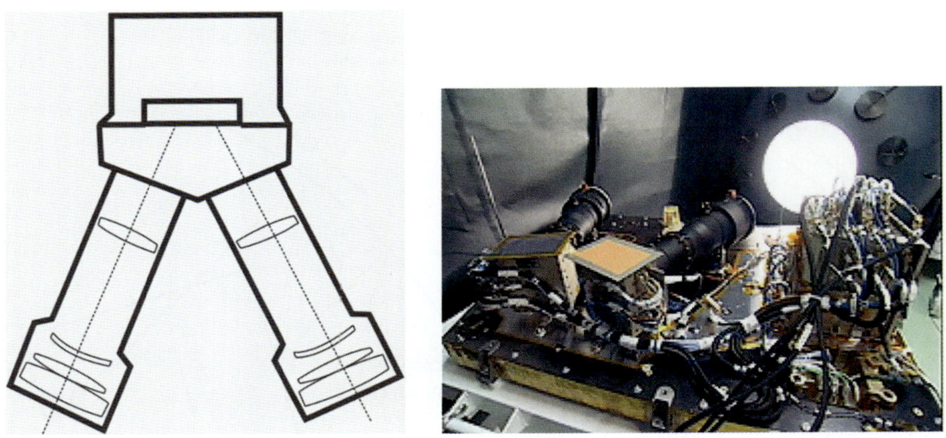

(a) *(b)*

Figure 4.15 SPOT-5 HRS: (a) principle of SPOT-5 HRS two optics, one sensor head; and (b) the SPOT HRS sensor. Credit: Spot Image

5 m in-orbit and 10 m across-orbit direction is generated. This includes the advantage of an improved height determination, which is dependent upon the pixel size in the orbit direction. Only the forward or the backward view can be taken, limiting the possible length of a stereo scene to 600 km (Figure 4.16). The HRS sensor has the same spectral range as the panchromatic SPOT band, leading also to problems with automatic image matching over forest areas.

HRS stereo pairs are not distributed by Spot Image; only the digital elevation models that based on it are sold as Reference3D (see Section 7.4.5). According to Spot

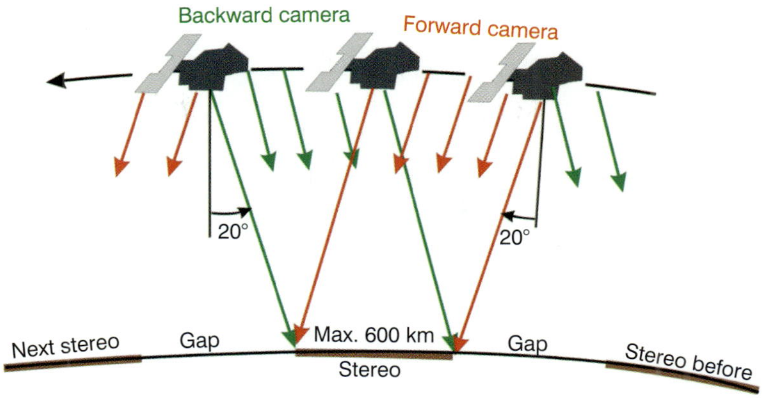

Figure 4.16 Imaging configuration of SPOT-5 HRS.

Image, the Reference3D with 1 arcsec sampling (31 m at the equator) is specified with 7 m standard deviation for the height and 10 m for the horizontal location. In forest areas, a gap-filling with the SRTM height model may occur (see Sections 1.4 and 7.4.2). An international test of SPOT HRS was made by the SPOT HRS study team of the International Society of Photogrammetry and Remote Sensing (ISPRS) in 2003 [Baudoin *et al.*, 2004].

4.3.3 MOMS

The German Modular Optoelectronic Multispectral Stereo Scanner MOMS-02 introduced the along-track stereo acquisition geometry for the first time on a civilian mission (see Section 2.7.2). MOMS-02 was flown on the Space Shuttle flight STS-55 in 1993 with 4.5 m panchromatic GSD and, as MOMS-2P, it was attached to the Russian Priroda module of the MIR space station. From the higher flying elevation of the space station, only 5.8 m GSD was reached. The MIR space station had problems with the power supply and MOMS-2P was not mounted in an optimal position, leading to a limited number of space images. But MOMS had a strong influence on the development of other stereo systems. The three cameras for the panchromatic channel, viewing to the nadir with 4.5 m GSD during the MOMS-02 mission and with 13.5 m GSD 21.4° ahead and 21.4° behind, enabled in general a stereoscopic coverage. The same location on the ground is imaged 20 sec and 40 sec later (Figure 4.17), enabling a 3D mapping, including the generation of digital elevation models. As mentioned before, the short time interval avoids disturbing changes in the object which have to be expected in the case of a stereo arrangement based on images taken from different orbits. The base-to-height ratio of 0.4 for the combination of nadir and fore and aft images and 0.8 for the combination of fore and aft images, guarantees a high quality of digital height models.

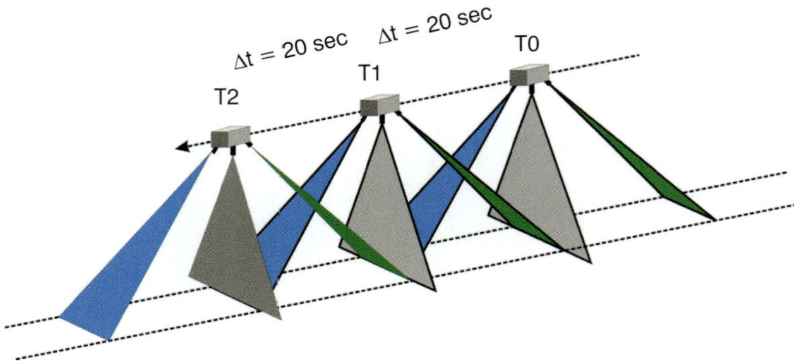

Figure 4.17 Stereoscopic coverage by MOMS with time differences of mission MOMS-02.

4.3.4 Japanese JERS and ASTER satellites

The Japanese programme for mapping started in 1992 with the Japanese Earth Resources Satellite 01 (JERS-1) [JERS1, 2011]. JERS-1 is an Earth Observation Satellite for covering the global land area for national land survey, as well as for other environmental applications. It was launched into a sun-synchronous sub-recurrent orbit at an altitude of 568 km, with a recurrent period of 44 days, by the H-I launch vehicle from the National Space Development Agency of Japan (NASDA) Tanegashima Space Center on 11 February 1992, and observed and collected data using a high performance synthetic aperture radar (SAR) and optical sensor (OPS). OPS can observe in seven bands ranging from the visible region to the short-wave infrared band, and is capable of stereoscopic observation by viewing ahead 15.3° from nadir in the near-infrared band.

The Japanese Advanced Spaceborne Thermal Emission and Reflection Radiometer, or ASTER [ASTER, 2011], is an imaging instrument mounted on TERRA, a US satellite launched in December 1999 as part of NASA's Earth Observing System. The primary objective of the ASTER mission is to obtain high spatial resolution images of the Earth in 14 spectral bands. ASTER consists of three different subsystems: the visible and near infrared (VNIR, 15 m GSD), the short-wave infrared (SWIR, 30 m GSD) and the thermal infrared (TIR, 90 m GSD). The VNIR subsystem consists of two independent telescopes, a backward- and a nadir-looking one, generating stereoscopic coverage (see Figure 4.18). The focal plane of the nadir telescope contains three CCD linear arrays (bands 1, 2, 3N), while the focal plane of the backward telescope has only one (band 3B).

The VNIR subsystem nadir-looking telescope is a reflecting-refracting improved Schmidt design. The focal plane of this telescope contains three 5000-pixel CCD-line arrays. The nadir- and backward-looking telescope pair are used for same-orbit stereo imaging and can be rotated as a unit ±24 degrees to provide extensive cross-track pointing capability. The VNIR subsystem backward-looking telescope is of the same design as the nadir telescope.

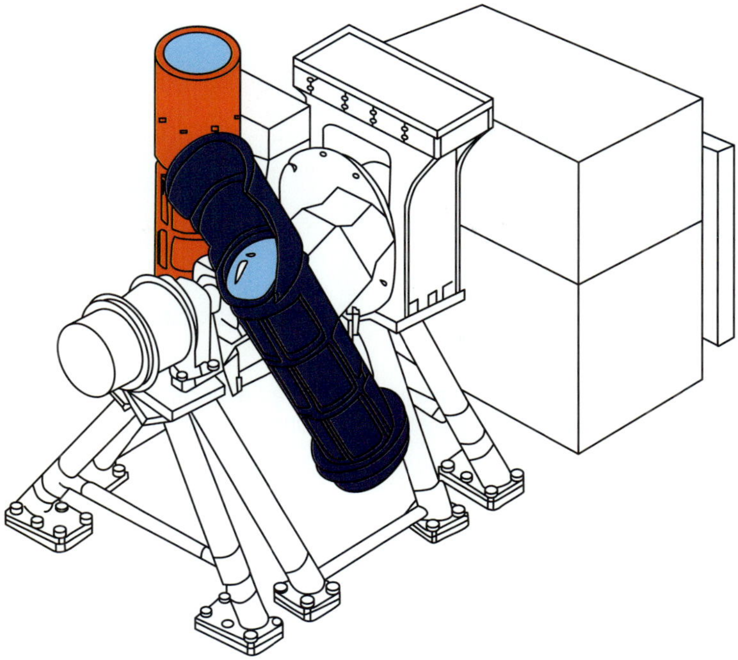

Figure 4.18 ASTER VNIR sensor: the backward-looking telescope is shown in red and the nadir-looking telescope is shown in dark blue. Credit: JAXA

The two near-infrared spectral bands (0.76–0.86 μm wavelength) 3N and 3B generate along-track stereo pairs, with an angle of convergence of about 27.7°. In this configuration it takes approximately 64 seconds to generate a stereo pair. The along-track image acquisition configuration has a strong advantage, in terms of radiometric variations, over a multi-date stereo data acquisition with across-track stereo. With the near-infrared band instead of panchromatic, conditions are optimal for a good radiometric response. A good object contrast is also given in forest areas in the near-infrared spectral range, causing no problems for automatic image matching.

The high resolution band is based on staggered CCD lines. The grey value level of both combined CCD lines is not improved by an actual radiometric calibration, causing image striping. Adjacent columns show a constant grey value shift with respect to each other, but a simple shift to the average grey value of neighbouring columns removes the striping completely (Figure 4.19). This striping has to be removed by the user.

4.3.5 ALOS PRISM and AVNIR-2

The Japanese space organisation JAXA launched the Advanced Land Observing Satellite (ALOS) in 2006. ALOS followed the Japanese Earth Resources Satellite-1 (JERS-1) and Advanced Earth Observing Satellite (ADEOS). Its main purpose is

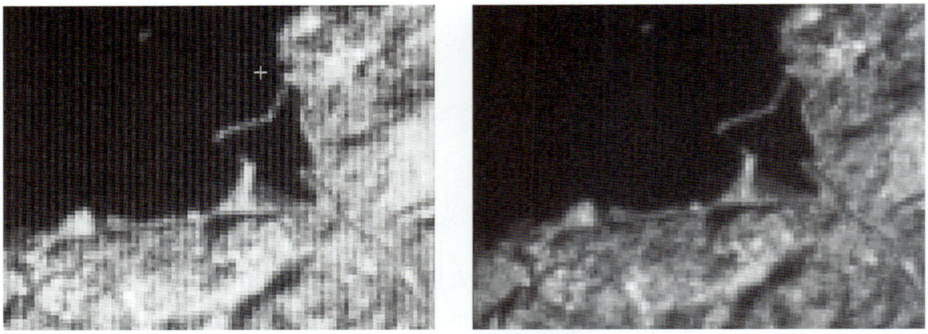

(a) (b)

Figure 4.19 Striping seen in ASTER images before and after removal: (a) original ASTER sub-image affected by striping; and (b) ASTER sub-image after striping correction.

land observation. ALOS, also named Daichi, was designed for cartography, regional observation, disaster monitoring and resource surveying. ALOS carries the Phased Array type L band Synthetic Aperture Radar (PALSAR) with a range resolution of 7 m, the Advanced Visible and Near-Infrared Radiometer type 2 (AVNIR-2) with 10 m GSD and the Panchromatic Remote-sensing Instrument for Stereo Mapping (PRISM), a combination of three cameras for stereoscopic coverage with 2.5 m GSD.

The 7000 pixels of the AVNIR-2 CCD lines cover a swath of 70 km. The location of the swath can be changed by viewing up to ±44° across the orbit – or 750 km on both sides of the orbit. This corresponds to a change of the GSD of 10 m in the nadir direction to up to 19.3 m GSD in the direction across the orbit. The spectral bands of AVNIR are shown in Table 4.4.

PRISM images the ground in the extended panchromatic range of 0.52 to 0.77 μm wavelength with three cameras – in the nadir, 24° forward and 24° backward directions (Figure 4.20). The base-to-height ratio of the possible stereo pair configurations is 0.5 for the combination of the nadir camera with one of the other two, or 1.0 for the combination of the forward-viewing and the backward-viewing cameras. The stereo models have a swath of 35 km width, guaranteed without influence of Earth rotation because of additional image parts of the inclined images. Only the nadir camera covers 70 km swath width.

As is common, ALOS PRISM does not have a solid CCD line, but a combination of six lines (Figure 4.21). Each single CCD has 4960 pixels; the combination

Table 4.4 Spectral range of AVNIR-2 bands

	Band 1	Band 2	Band 3	Band 4
Wavelength (μm)	0.42–0.50	0.52–0.60	0.61–0.69	0.76–0.89
Colour	blue	green	red	near infrared

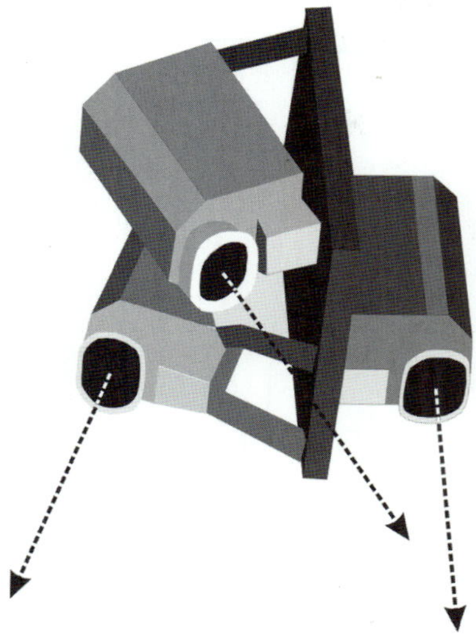

Figure 4.20 Configuration of ALOS PRISM cameras.

Figure 4.21 The arrangement of CCD lines in the focal plane of PRISM.

covers the whole swath width with a small overlap of neighbouring CCD lines. For fitting adjacent CCD lines together, the small time delay of imaging different locations in the focal plane orbit direction has to be respected (Figure 4.22). In theory, a correct fit is only possible for a specified object height level, but in most cases the misfit of objects not located in the height level of merging the sub-images can be neglected.

Three different data products are available: level 1A are raw data; level 1B1 have a radiometric correction and are corrected according to the sensor calibration; and georeferenced level 1B2 are projected to a plane of constant height. Level 1A data are recommended only for scientific purposes because the user has to make the geometric correction of the six individual sub-images, including the inner orientation, and has to stitch the sub-images from the six CCD lines together to form a solid image [Kocaman and Gruen, 2007].

The combination of three images taken from different directions leads to an over-determination of 3D object points. This includes the advantage of blunder detection for digital elevation models determined by automatic image matching. The ALOS project is not directly commercial; images are distributed for a low price to developing countries, and sometimes free of charge. On the other hand, the image quality is variable and refinement often has to be made by the user.

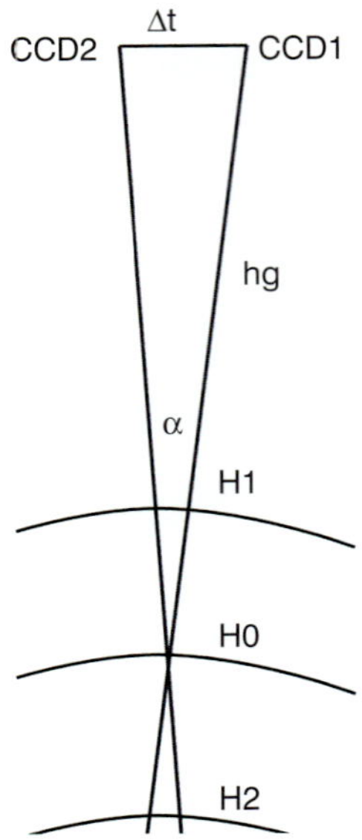

Figure 4.22 The misfit of neighbouring CCD lines at slightly different locations in the focal plane orbit direction, depending on height difference from the reference height.

4.3.6 Indian programme

4.3.6.1 IRS-1C, IRS-1D, Resourcesat

The Indian Space Research Organisation (ISRO) has launched a series of optical satellites. From the launch of IRS-1C in 1995 up to the end of 1999, IRS-1C and IRS-1D, which were nearly identical satellites with 5.8 m GSD, delivered the highest resolution optical space images. Similar to SPOT, inclined images can be taken from a panchromatic camera rotated to the side. In addition, the LISS-III camera provides green (0.52–0.59 μm), red (0.62–0.68 μm) and near-infrared (0.77–0.86 μm wavelength) bands with 23.5 m GSD. A short-wave infrared (1.55–1.70 μm) band with 70.5 m GSD improves the possibility of object classification. LISS-III takes images only in the nadir direction with a 141 km swath. The Wide Field Sensor (WiFS), with two cameras operating in the red band (0.62–0.68 μm) and two near-infrared cameras (0.77–0.86 μm), covers a swath of 810 km with 188 m GSD. Because of the image compression required for data transfer from the satellite to the ground stations, the panchromatic images are limited to 6 bits (64 grey values), while LISS-III and WiFS provide images of 7 bits (128 grey values).

With IRS Resourcesat-1, also named IRS-P6, when the camera operates in the multispectral mode, the colour information from the LISS-IV camera has been available with 5.8 m GSD over a swath of 23.9 km since 2003. In a panchromatic mode, a swath of 70 km is covered. The bands have the same spectral range as the LISS-III camera. 7-bit grey values are optimally selected from the original 10-bit images. In addition, Resourcesat carries a LISS-III camera, which is nearly identical to the LISS-III camera on IRS-1C and IRS-1D. The Advanced Wide Field Sensor (AWiFS) camera improved the ground resolution from 50 m to 70 m GSD over a 737 km swath. This group of satellites from ISRO has a large imaging capacity, resulting in these images being used worldwide.

4.3.6.2 Cartosat 1 (IRS-P5)

ISRO has the most complete programme of Earth observation satellites, ranging from very high to low resolution systems. It has an intense programme for the development of improved optical and radar satellites and also replaces old systems. Between 1995 and 2010, ISRO successfully launched 11 Earth observation satellites and this programme will continue. Cartosat-1, also named IRS-P5, is a satellite for the generation of stereo pairs with 12 000 pixels of 2.5 m GSD in the extended panchromatic wavelength range of 0.50 to 0.85 μm. The spectral range extended to near-infrared has significant advantages for forest areas, where a satisfactory object contrast can be expected [Büyüksalih and Jacobsen, 2008].

The combination of the forward camera, with 26° nadir angle, and the backward camera, with –5° nadir angle, generates stereo models with a base-to-height ratio of 0.6, guaranteeing a good vertical accuracy of the stereo models for images taken with a time interval of 52 seconds. The on-board solid-state recording enables a global coverage. A systematic acquisition to cover whole continents continues. The Cartosat-1 images are supplied with rational polynomial coefficients (RPC) (see Section 6.3.2.2), including the orientation based on direct sensor orientation of the satellite. The absolute location accuracy without the influence of terrain relief was estimated at 220 m with no control points by Krishnaswamy and Kalyanaraman (2002). With the exception of the initial scenes after launch, it is now between 100 and 250 m. With four control points, this can be improved to less than 2 m [Jacobsen, 2006].

The satellite can roll to the side to also cover areas not located directly below the orbit (Figure 4.23). With a 23° roll angle the revisit capacity is five days and large areas can be covered with permanent imaging. This is not the same for a satellite generating a stereo model from different orbits – in such a case the length of the stereoscopic covered area is limited by the convergence of the areas imaged from neighbouring orbits (Figure 4.1a). By changing the satellite roll instead of a stereoscopic coverage, a single coverage with 56 km swath is possible.

4.3.7 Agile satellites in the 1–15 m range

Since 1999 a number of nations previously without a space programme have launched satellites with sensors designed for mapping. These have been agile satellites with a single telescope which can be pointed in any direction.

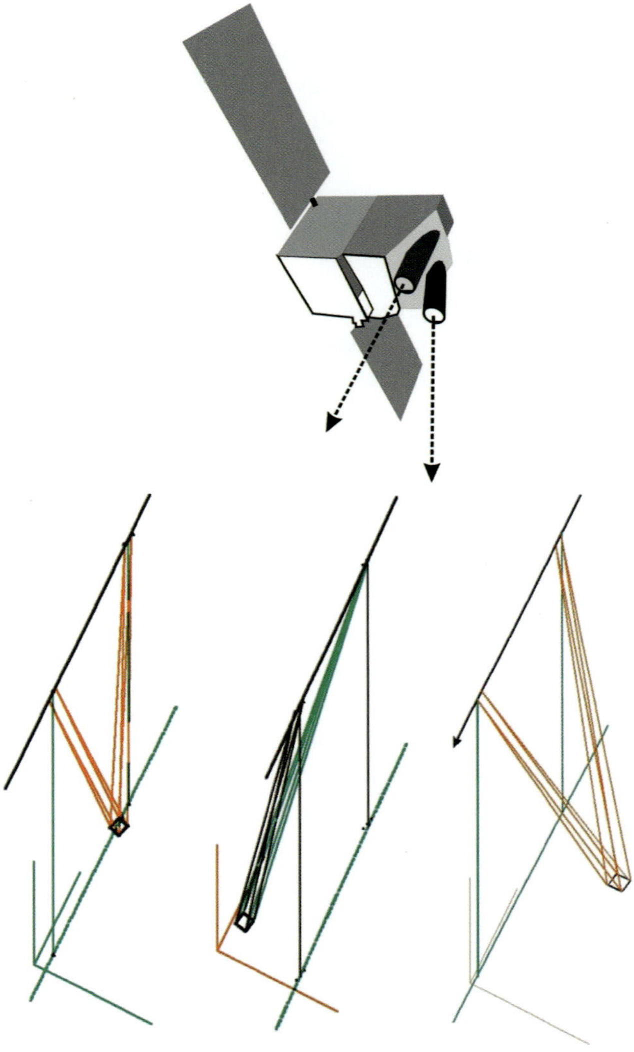

Figure 4.23 Sketch of Cartosat and imaging configurations for different roll angles.

4.3.7.1 KOMPSAT-1

Like India, South Korea has a series of optical satellites. KOMPSAT-1, from the South Korean Space Agency KARI, was launched in 1999. Its panchromatic electro-optical camera (0.51–0.73 μm wavelength), with 6.6 m GSD, has a swath of 17 km. With the OSMI camera, five spectral bands cover a swath of 800 km with 1 km GSD. KOMPSAT-1 is a satellite intended more for testing purposes than for production. Nevertheless the panchromatic images have a high quality and the spectral range of the panchromatic band is extended to near infrared. So a good object contrast is guaranteed also over forest areas. The whole satellite rotates to obtain stereoscopic coverage.

Figure 4.24 The arrangement of CCD lines in the focal plane of EROS A.

4.3.7.2 EROS A

EROS A, launched by ImageSat International in 2000, is a lightweight satellite of 250 kg. ImageSat International is officially located in the Netherlands Antilles, but it is dominated by Israeli shareholders. The satellite was built in Israel. Although the low weight reduces the cost, it causes some limitations of the system, such as having only a panchromatic band with 1.8 m GSD. With over-sampling, a 1 m GSD can be achieved for a swath limited to 10 km. EROS A is an agile satellite with a possibility of changing the view direction very quickly by means of powerful reaction wheels. During imaging, the satellite can have a nadir angle of up to 45° in any direction. Because of a sampling rate limited to 750 lines/sec, an asynchronous imaging mode with a slow-down factor of 5.2 (Figure 4.5) has to be used. A smaller slow-down factor leads to larger pixels in the orbit direction, which have to be interpolated to a smaller GSD. This, of course, causes a degradation of the image quality.

Like several other satellites, EROS A does not have a single CCD line; it has a combination of four CCD lines (Figure 4.24), which are joined together to form a unique line with 7043 pixels, covering a swath of 13 km when viewing in the nadir direction. The separate sub-images of the four CCD lines are handled as level "0A" images. EROS A is able to generate stereo pairs from the same orbit. Only a direct downlink from the satellite to the ground station is possible, requiring a network of ground stations with a broadcasting radius of approximately 2000 km. In addition, the imaging capacity is limited by the slow-down factor, so EROS A cannot cover large parts of the world like SPOT-5.

EROS A takes images up to an incidence angle of 45°. In the case of 45° along track, the projected pixel size in the track direction is 3.6 m instead of 1.8 m (Figure 4.3). The GSD in the track direction is still 1.8 m because of the slow-down factor, leading to an over-sampling of 50%. A 45° incidence angle across the view direction leads to 3.6 m GSD in the direction perpendicular to the orbit. Nevertheless, level 1B type images are distributed with 1.8 m GSD, with limited image quality in most cases. The large slow-down factor during imaging and some over-sampling also reduce the image quality of level 1A type scenes. A factor of approximately 1.5 for the effective GSD has been found by edge analysis. So the effective GSD of level 1A type scenes, corresponding to the information contents of the images, is more realistic with 1.5×1.8 m = 2.7 m GSD (see also Figure 4.6). This is easily confirmed by image inspection.

4.3.7.3 FORMOSAT-2

ROCSAT-2, later renamed FORMOSAT-2, was launched in 2004 by the Taiwanese China National Space Programme Office (NSPO). It was built by the European company EADS Astrium SAS. With a weight of 750 kg it still belongs to the category of small satellites. The extended panchromatic band covering the spectral range of 0.45–0.90 μm wavelength with 2 m GSD has a swath of 24 km, as do the four multispectral channels having 8 m GSD with blue (0.45–0.52 μm), green (0.52–0.60 μm), red (0.63–0.69 μm) and near infrared (0.76–0.90 μm). The large mirror optics with an aperture of 60 cm enables the 2 m GSD to be obtained without the need of TDI. Based on four reaction wheels, a free-body pointing of the planned object area is possible with a range of ±45° in any direction, providing a daily revisit capability. An on-board storage capacity of 41 Gb permits worldwide coverage. The direct sensor orientation guarantees a positioning of 70 m without control points. FORMOSAT-2's orbit is designed to cover Taiwan every day. The satellite completes exactly 14 orbits per day, which means it does not give complete coverage of the Earth. The FORMOSAT-2 images are distributed by Spot Image.

4.3.7.4 Beijing 1

Surrey Satellite Technology Ltd. (SSTL) [SSTL, 2011] built the Beijing 1 satellite under the working title DMC+4, based on a contract with Beijing Landview Mapping Information Technology Co (BLMIT). It was launched in October 2005 and the first attempt by this group to commercialise data services using Earth observation satellites. A panchromatic camera with 4 m GSD and a swath of 24 km was added to a standard DMC satellite from SSTL, which had 32 m GSD multispectral imaging capability, with a swath of 640 kilometres. The 166 kg spacecraft provides a large data storage capacity, with solid-state storage augmented by hard drives modified for use in space, allowing the instrument to map strips of 4000 km in length. The transmission of the images to the ground station is supported by on-board programmable data compression.

4.3.7.5 TopSat

TopSat was launched in 2005 in a cooperation of the UK Ministry of Defence and the British National Space Centre (BNSC) in the frame of the small-satellite initiative. TopSat demonstrates the enhanced capability of small satellites. Led by QinetiQ, UK, the space bus was constructed by SSTL – a company that is very active in the construction of small satellites for several countries. TopSat, like the group of small satellites of the Disaster Mapping Constellation (DMC), organised by SSTL, is included in the International Charter "Space and Major Disasters", a joint initiative by global space agencies for making actual images of disaster areas available within a short time [Chander *et al.*, 2008].

TopSat has a panchromatic channel with 2.8 m GSD and three multispectral channels in the visible spectral range with 5.6 m GSD. It weights only 130 kg and was not constructed for covering large areas. Only a direct downlink of images is possible, but this can also be done to a mobile ground station. The focal length of the folded mirror is 1.68 m, to fit into the compact, lightweight satellite construction. This requires a pixel size of 7 μm for the panchromatic channel. Standard Kodak TDI arrays have a pixel size of 12 μm, so a standard CCD linear array is used, requiring an asynchronous imaging mode with a slow-down factor of up to eight to obtain a satisfactory signal-to-noise ratio. The imaging capacity of TopSat is limited to one image per orbit. Four panchromatic and multispectral images can be stored in the satellite [Wallace and Parker, 2001]. TopSat has become a high quality imaging system, with direct user access from transportable ground facilities.

4.3.7.6 CBERS-2B

The China-Brazil Earth Resources Satellite CBERS-2B, launched in 2007, is based on a partnership between the National Institute for Space Research (INPE) in Brazil and the Chinese Academy of Space Technology. CBERS-2B carries a panchromatic camera that images with 2.4 m GSD, a multispectral camera with 20 m GSD in four colour and one panchromatic bands, and a wide field camera with 260 m GSD in two bands (Table 4.5). The high resolution panchromatic camera replaced the infrared scanning camera on board CBERS-1 and CBERS-2.

The multispectral camera uses a rotating mirror with up to ±32° nadir angle across the orbit, similar to SPOT. This leads to a revisit time of three days. Variations of CBERS 1, 2 and 2B include an on-board storage in order to also image areas outside the range of the ground stations. Images of CBERS satellites are distributed free of charge in Brazil, China and Africa.

4.3.7.7 THEOS

Thailand's space agency, GISTDA, launched the THEOS satellite in October 2008. THEOS, like FORMOSAT-2, was built by the European company EADS Astrium SAS. The panchromatic band covers, with 2 m GSD, a swath of 22 km; and the multispectral band covers, with 15 m GSD, a swath of 90 km. This is an unusual relationship between the panchromatic and the multispectral resolution. Nearly the same spectral

Table 4.5 Characteristics of CBERS-2B cameras

Camera	Bands (μm)	GSD (m)	Swath (km)
Panchromatic (HRC)	0.51–0.73	2.5	27
Multispectral (CCD)			
blue	0.45–0.52	20	113
green	0.52–0.59	20	113
red	0.63–0.69	20	113
NIR	0.77–0.89	20	113
pan	0.51–0.73	20	113
Wide field imager (WFI)	0.63–0.69	260	890
	0.77–0.89	260	890

sensitivity as FORMOSAT-2 is given. By body pointing, the agile satellite is able to generate stereo models in the same track. Images can be taken with nadir angles of ±50° in any direction, but usually limited to ±30°. An on-board memory enables imaging outside the range of ground stations.

4.3.7.8 RapidEye

Most optical satellites are made for dual use – military reconnaissance and civilian application – or they are civilian governmental systems. An exception is the private German company RapidEye, which launched a group of five satellites in August 2008, without government financial support or an advance order from the military. The images are distributed by MacDonald Dettwiler, Canada.

The bus platform of the satellites was constructed by SSTL, while the payload comes from Jena-Optronik in Germany. The satellites have a weight of only 150 kg each and a proposed lifetime of seven years. With 12 000 CCD elements and 6.5 m GSD, the swath is 78 km for nadir viewing. The main purpose of the group of satellites is the support of insurance and food companies, farmers, governments and others. To do this, a sufficient spectral resolution is given with five spectral bands: blue (0.44–0.51 μm), green (0.52–0.59 μm), red (0.64–0.685 μm), red edge (0.69–0.73 μm) and near infrared (0.76–0.85 μm). The constellation will be able to cover any specified area on the entire globe every day. This is only possible with the satellites rolling up to 45° across the orbit.

With modern technology, such a group of small satellites today can be very powerful. The components are getting smaller, reducing the weight as well as the power consumption and enabling the launch of all five satellites with just one rocket.

4.3.8 Small satellites

In addition to the optical satellites mentioned above, there exists another group of small satellites. These are not specifically made for commercial application, so their imaging capacity is limited. Furthermore, it is difficult to order scenes for a specified area, but nevertheless these small satellites are becoming more and more powerful. By definition, the RapidEye satellites also belong to this group, but they are entirely commercial.

UoSAT-12 belongs to the more experimental small satellites. It was built by SSTL, as were several other small satellites. With 10 m GSD panchromatic it boasts the same GSD as SPOT 1–4, but does not have the same imaging flexibility. Instead of a CCD line, UoSAT-12 is equipped with CCD arrays covering 10 km × 10 km panchromatic and 30 km × 60 km multispectral.

SunSat from SunSpace, South Africa, active from 2000 to 2002, had a panchromatic channel with 15 m GSD and 52 km swath. The satellite was limited to 64 kg. BilSat-1 from Tübitak, Turkey, constructed by SSTL and active from 2003 to 2006, was equipped with CCD arrays like UoSAT-12. A 12.6 m GSD panchromatic band and four multispectral bands with 28 m GSD were in use, but it was difficult to procure images taken by this satellite over specified areas. The small satellite KITSAT-3 with 15 m green, red and near-infrared images, covering a swath of 50 km, was also more experimental and South Korea's first step in a successful series of optical satellites. Turkey is also now working on a very high resolution satellite.

4.4 Space photography

4.4.1 CORONA

During the Cold War the USA and the USSR developed reconnaissance satellites for military purposes. The USA created the CORONA camera system, with different versions from 1960 onwards up until 1972 (Table 4.6), when the film-based satellite systems were replaced by digital cameras.

In 1995 the CORONA images were declassified and are now available as scanned photographs. Many of these images are now over 30 years old and cannot be used for map updating, but for several applications the generation of digital elevation models (DEM)and the old object information are important. The height models may be useful in areas with unchanging surfaces, where they can be used together with actual space images for the computation of orthoimages, or in areas with strong erosion as a reference for the detection of changes. Moreover, the CORONA images are excellent for archaeological purposes. The stereoscopic KH-4B camera was the most often used system (Figure 4.25). Nowadays the images can be ordered via the Internet for a handling fee.

With the exception of the KH-5, CORONA used panoramic cameras, scanning from one side to the other as shown in Figure 4.26. The field of view, however, is smaller than that shown in the figure, so the geometric difference between a perspective and a CORONA panoramic image is limited. The panoramic deformation can be used as a pre-correction, and the dynamic change of the geometry can be determined by self-calibration with certain additional parameters [Schneider *et al.*, 2001].

Table 4.6 CORONA camera systems [McDonald, 1997]

System	Period used	No. of cameras	Focal length (inches)	Format (inches)	Resolution (ft)
KH-1	08/60	1	24	2.2 × 30	40
KH-2	12/60–7/61	1	24	2.2 × 30	30
KH-3	8/61–12/61	1	24	2.2 × 30	25
KH-4	2/62–12/63	2	24	2.2 × 30	25
KH-4A	8/63–10/69	2	24	2.2 × 30	9
KH-4B	9/62–5/72	2	24	2.2 × 30	6
KH-5	5/62–7/64	1	3	4.5 × 4.5	460
KH-6	7/63–8/63	1	66	4.5 × 25	2

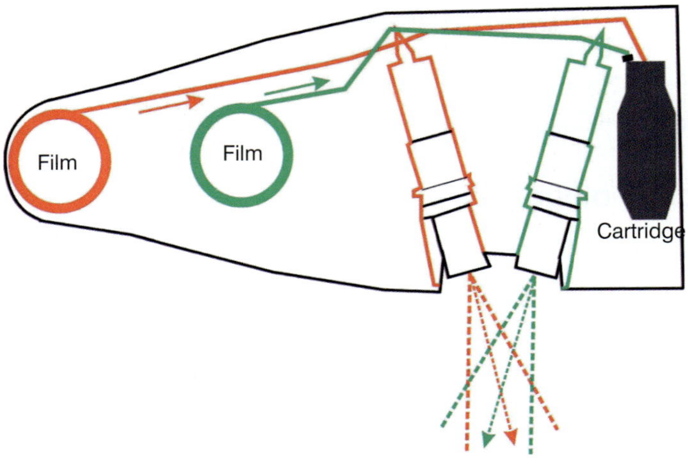

Figure 4.25 Basic principle of the CORONA KH-4B camera configuration.

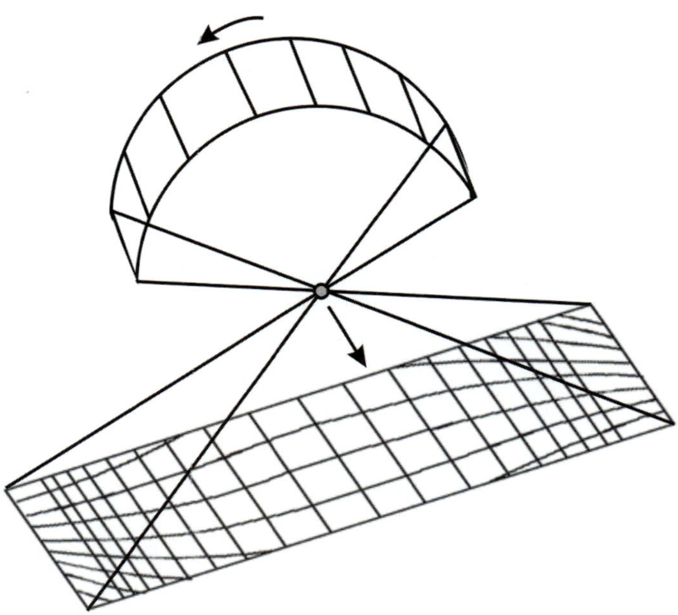

Figure 4.26 Geometric principle of a panoramic camera, showing the deformation of a square grid in object space, imaged to the photos.

DEMs can be generated based on CORONA KH-4B stereo pairs. A DEM of the Chinese loess plateau (Figure 4.27), heavily influenced by erosion, has been generated, with an estimated absolute standard deviation of the height of approximately 14 m [Schneider *et al.*, 2001]. The relative accuracy is around 5 m.

Figure 4.27 A grey value scaled digital elevation model of the Chinese loess plateau based on a CORONA KH-4B stereo model [Schneider *et al.*, 2001].

4.4.2 European and US photography

The German metric camera and the large format camera from the USA are discussed in Section 2.7.1.

4.4.3 Soviet photography

The USSR, now Russia, used a system of film cameras in space as shown in Table 4.7. The photographic cameras were used from low orbits of approximately 220 km, but also from the MIR space station, operating at a height of 350 km.

The ground resolution as given by Soyuzcarta has been shown to be too optimistic in relation to digital space images. Nevertheless, with KFA1000, KFA3000 and panoramic KVR1000 photographs, quite a bit more object detail has been identified than with SPOT-4 images. The evaluation of height models based on TK350 and ASTER (15 m GSD) models in the same area resulted in slightly better accuracy with ASTER images. In forest areas the TK350 matching failed in a large percentage of cases

Table 4.7 Technical data of Russian photographic space cameras

Sensor	KFA 1000	MK4	KFA 3000	KVR 1000	TK350
Focal length (mm)	1000	300	3000	1000	350
Image size (mm)	300 × 300	180 × 180	300 × 300	180 × (180)	300 × 450
Orbiting height (km)	220/350	220/350	220/350	220/350	220/350
Area covered	66/105	132/210	22/35	40 × 180 64 × 280	200 × 300 310 × 470
Ground resolution (m/lp)	5–10	10–15	2–5	2–5	10–15
Base-to-height ratio	0.12	4.2	no stereo	no stereo	0.55

[Büyüksalih *et al.*, 2003]. In addition, the delivered TK350 photographs had a lot of scratches and were very grainy: the high price of TK350 photographs was not justified compared to the cheaper ASTER stereo scenes. Space photographs are no longer taken today, and only old photographs can be used.

References

ASTER, 2011. http://asterweb.jpl.nasa.gov/instrument.asp (accessed September 2011).

Baudoin, A., Schroeder, M., Valorge, C., Bernard, M. and Rudowski, V., 2004. The HRS-SAP initiative: A scientific assessment of the High Resolution Stereoscopic instrument on board of SPOT 5 by ISPRS investigators. *International Archives of Photogrammetry, Remote Sensing and Spatial Information Sciences*, 35(B1), 372–378.

Büyüksalih, G., Kocak, M.G., Oruc, M., Akcin, H. and Jacobsen, K., 2003. DEM generation by ASTER and TK350. Joint Workshop High Resolution Mapping from Space, September, Hannover, Germany, http://www.ipi.uni-hannover.de (accessed September 2011).

Büyüksalih, G. and Jacobsen, K., 2008. DSM generation with high resolution space imagery over mountainous forest area, *International Archives of Photogrammetry, Remote Sensing and Spatial Information Sciences*, 37(B1), 865–871.

Chander, G., Choate, M. and Scarramuza, P., 2008. Geometric assessment of TopSat and Beijing-1 data. Civil Commercial Imagery Workshop, Fairfax, VA, USA, http://calval.cr.usgs.gov/JACIE_files/JACIE08/2008_JACIE_DVD/Wednesday_PM/Chander_TopSat-BEIJING-1.pdf (accessed January 2011).

Jacobsen K., 2006. ISPRS-ISRO Cartosat-1 scientific assessment programme (C-SAP) technical report – test areas Mausanne and Warsaw. *International Archives of Photogrammetry, Remote Sensing and Spatial Information Sciences*, 36(4):1052–1056.

JERS1, 2011. http://www.eorc.jaxa.jp/JERS-1 (accessed September 2011).

Kocaman, S. and Gruen A., 2007. Orientation and calibration of ALOS/PRISM imagery. *International Archives of Photogrammetry, Remote Sensing and Spatial Information Sciences*, 36(1/W51), http://www.ipi.uni-hannover.de (accessed January 2011).

Krishnaswamy, M. and Kalyanaraman, S., 2002. Indian remote sensing satellite Cartosat-1: Technical features and data products. GIS Development, http://www.gisdevelopment.net/technology/rs/techrs023.htm (accessed February 2011).

McDonald, R.A. (Ed.), 1997. *CORONA Between the Sun & the Earth.* American Society for Photogrammetry and Remote Sensing, Bethesda, MD, USA.

Schneider, T., Seitz, R., Förster, B. and Jacobsen, K., 2001. Remote sensing based parameter extraction for erosion control purposes in the Loess plateau of China. High Resolution Mapping from Space 2001, Leibniz University Hannover, http://www.ipi.uni-hannover.de/uploads/tx_tkpublikationen/jac_CORONA.pdf (accessed February 2011).

SPOT5, 2011. http://spot5.cnes.fr/gb/satellite/satellite.htm (accessed September 2011).

SSTL, 2011. http://www.sstl.co.uk (accessed September 2011).

Wallace, K.M. and Parker, I., 2001. The Topsat satellite. IEEE International Geoscience and Remote Sensing Symposium, Sydney, Australia, 3246–3248, http://ieeexplore.ieee.org/xpl/freeabs_all.jsp?arnumber=978317 (accessed February 2011).

Sensors with a GSD of 1 m or less | 5

Karsten Jacobsen

5.1 General information

The number of very high resolution satellites is growing all the time. In addition, the ground resolution is improving and will continue to improve in the future. The construction of GeoEye-2, with planned 25 cm GSD, has started, as well as the construction of Cartosat-3, with 33 cm GSD. At the end of 2010, optical space images with 1 m GSD or less are available from five countries (USA, Israel, India, Russia and South Korea) for civilian purposes and more countries will make an entrance in the coming years. The technological knowledge for this development is available in a group of countries; in addition several components, or even whole satellites, can be bought off the shelf. A competitive market exists for launching satellites. There was a delay in starting with very high resolution satellites, as a result of the failure of IKONOS-1 in April 1999 and QuickBird-1 in 2000, but the launch and the satellite components have become safer.

Figure 5.1 gives an overview of the launch dates and GSDs of the panchromatic channels of the high resolution optical satellites. The breakthrough for topographic mapping with satellite images and new applications was at the end of 1999, with the successful launch of IKONOS-2. Thirteen satellites with 1 m GSD or less, launched before the end of 2010, are included in Figure 5.1. Not all of them deliver images to the competitive market of high resolution space images, and the imaging capacity of the satellites is quite different, but it is easier to get actual high resolution space images from a specified area today than it was in the past.

The construction of sensors generating images with a GSD of 1 m or less is more difficult than for 5 m or larger GSD. Without over-sampling of the ground pixels, the exposure time for one pixel depends on the satellite footprint speed (Figure 5.2) if the view direction against the orbit is not changed during imaging by an asynchronous imaging mode (see Section 4.2).

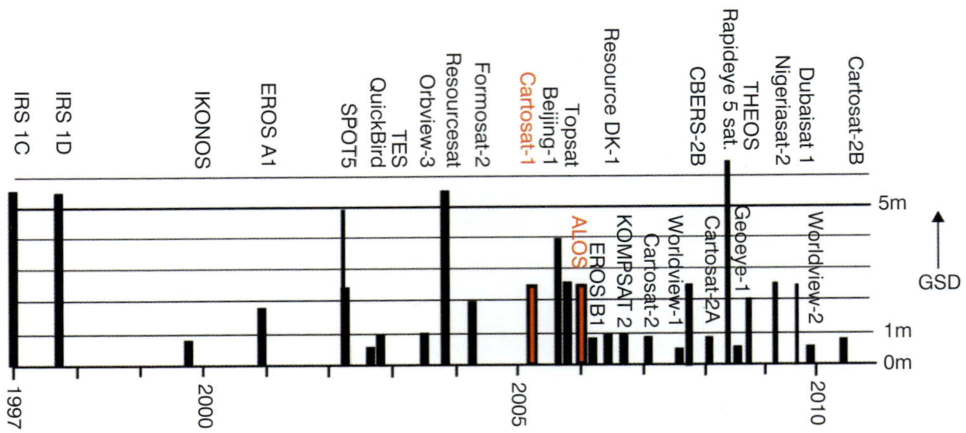

Figure 5.1 Overview of high and very high resolution optical satellites as a function of launch date and GSD for panchromatic bands.

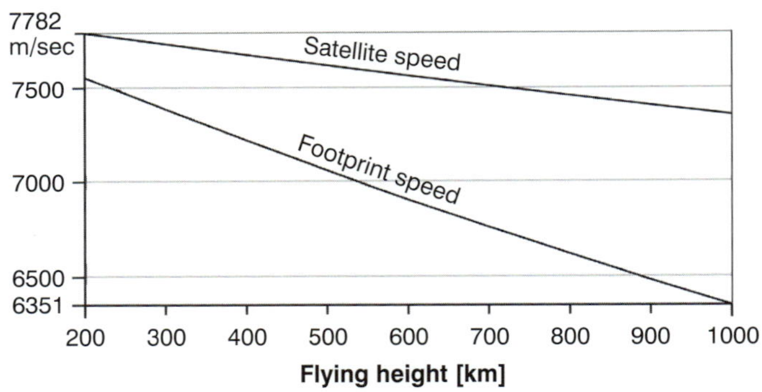

Figure 5.2 Satellite speed as a function of the flying height for a circular orbit with Earth radius = 6366 km.

If the footprint speed is approximately 7000 m s^{-1}, an exposure time of only 0.14 msec is available for 1 m GSD. Even with a large camera aperture, an acceptable signal-to-noise ratio cannot be achieved by imaging with a single CCD line. A transfer delay and integration system (TDI) (Figure 4.6), an asynchronous imaging mode (Figure 4.5) or staggered CCD lines (Figure 4.4), or a combination of these, is required. An asynchronous imaging mode reduces the imaging capacity because of the increased exposure time. For high imaging capacity a TDI system is required, but for high radiometric quality the number of stages is limited for inclined images.

From a lower altitude the required angular resolution is not as much as from higher levels. On the other hand, an increased flying height improves the revisit time and the time for changing from one area to another. The slewing speed, or rate of change of pointing direction, is based on reaction wheels; or control moment gyros can be used to increase the slewing speed.

Key components for the comparison of optical satellites are ground resolution, imaging capacity (Table 5.1), revisit time, agility and accuracy, and the financial considerations and conditions of provision. The imaging capacity is determined by the swath width, but more importantly by the internal storage capacity and downlink possibilities. In any case the images are heavily compressed before internal storage and downlink to the receiving station, which may happen via a relay satellite. Agility of the sensor is an important factor. For example, QuickBird-2 is able to generate stereoscopic coverage within the same track, but the lower agility (Table 5.1 and Figure 5.3) does not allow additional image collection between the first and the second image of the same area; the whole time is required for the satellite rotation. This makes stereo imaging with QuickBird-2 uneconomic. This changed with the very fast rotation capability of WorldView and GeoEye, and the generation of two different stereo pairs during the same part of a track, or the collection of a group of single scenes with different locations, was possible.

With growing competition the conditions for purchasing images have been improved. Based on large archives, images can be bought directly or can be ordered from different organisations. The distribution system is now operational, whereas it was not initially for some of the very high resolution satellites.

All high resolution optical satellites use body pointing for imaging, that is the whole satellite is rotated to the target. This leads to single optical instruments on one satellite. The satellites with the highest ground resolution resemble an optical instrument with some small additional components on it. The satellite agility is also used for the imaging itself. The scene on the ground is not following the orbit; by permanent change of the

Table 5.1 Comparison of imaging capacity (McGill 2005 and information from satellite companies)

Satellite	Agility – slew over 300 km (sec)	Collection rate (km^2/day)	Remarks
IKONOS-2	25	150 000	
QuickBird-2	62	135 000	Asynchronous mode
OrbView-3	31	51 700	Pan and MS not together
KOMPSAT-2		1 700 000 20 minutes/orbit	
WorldView-1	10	750 000 (pan only)	
GeoEye-1	25	350 000 (pan+MS) 700 000 (pan)	
WorldView-2	9	975 000 (pan)	8 spectral bands
Cartosat-2		528 000	

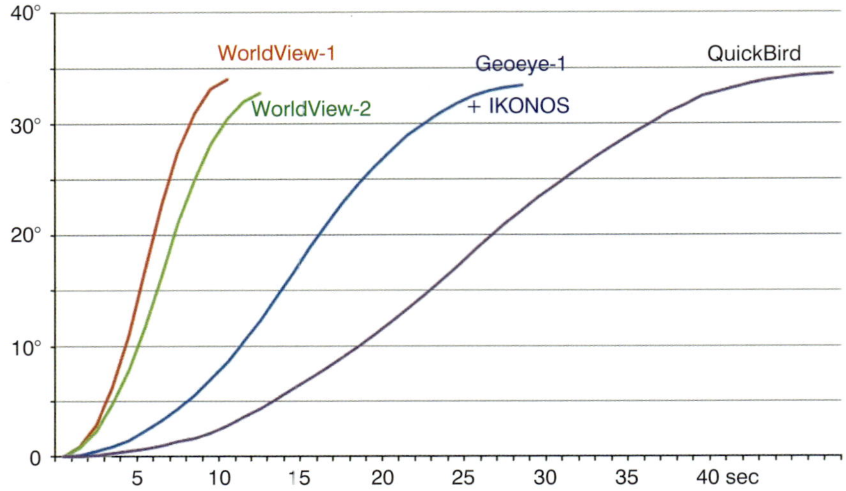

Figure 5.3 Slewing times of body-pointing satellites.

satellite attitude most scenes are oriented to the ground coordinate system, but they may have any orientation. In particular stereoscopic pairs are often scanned from east to west, and vice versa (see Figure 5.5) after rotating the satellite approximately 80° in yaw. This rotation during imaging does not influence the accuracy of georeferencing.

5.2 Sensors

5.2.1 Introduction

Table 5.2 shows current satellites having optical sensors with GSD ≤ 1 m. All these satellites are dual-use systems: the majority of funding comes from military reconnaissance, but the free capacity is available for civilian use. Optical reconnaissance satellites limited to military or internal national use are not included here. For example, France (with Helios) and Japan have optical satellites with a ground resolution belonging to this group. The USA, Russia and Israel also have additional very high resolution optical systems for intelligence collection. In addition, synthetic aperture radar (SAR) satellites with 1 m GSD are active, such as the civilian German TerraSAR-X, the military German SAR Lupe and the dual-use Italian COSMO-SkyMed.

SAR images with 1.0 m GSD cannot be compared directly with optical images. The information content – the amount of detail that can be identified – is lower for SAR than for optical images. The interpretation of SAR images is difficult and influenced by speckle and, especially in built-up areas, by layover and corner reflection. In mountainous areas the layover destroys a lot of information. On the other hand, radar can view through clouds and digital elevation models can be generated by interferometric SAR (InSAR). Using differential InSAR, changes in an object's height of up to approximately 1 cm can be mapped if the object is not disturbed by vegetation.

Table 5.2 Current sensors and satellites with GSD ≤ 1 m

Sensor or satellite	Agency Country launch date	Platform altitude (km)	GSD pan/Ms (m)	Swath width at nadir (km)	Channels wave-length (µm)	Quantisation (bits)
IKONOS-2	GeoEye Inc. USA 1999	681	0.82/3.28	11.3	pan 0.45–0.90 MS: 0.45–0.53, 0.52–0.61, 0.64–0.72, 0.77–0.88	11
QuickBird	DigitalGlobe USA 2001	450	0.61/2.44	16.5	pan 0.45–0.90 MS: 0.45–0.52, 0.52–0.60, 0.63–0.69, 0.76–0.90	11
OrbView-3	GeoEye Inc. USA 2003	470	1/4 pan and MS not together	8	pan 0.45–0.90 MS: 0.45–0.52, 0.52–0.60, 0.625–0.695, 0.76–0.90	11
EROS B	ImageSat Int. N. Antilles (Israel) 2006	508	0.7/–	7	pan 0.5–0.9	10
KOMPSAT-2	KARI South Korea 2006	685	1/4	15	pan 0.50–0.92 MS: 0.45–0.52, 0.52–0.60, 0.63–0.69, 0.76–0.90	10

(Continued)

Table 5.2 (cont.) Current sensors and satellites with GSD ≤ 1 m

Sensor or satellite	Agency Country launch date	Platform altitude (km)	GSD pan/Ms (m)	Swath width at nadir (km)	Channels wave-length (μm)	Quantisation (bits)
Geoton-L1 Resurs-DK 1	Roscosmos (Roshydromet) Russia 2006	330–585 elliptical orbit	1/3 at 350 km	30	pan 0.58–0.8 MS: 0.5–0.6, 0.6–0.7, 0.7–0.8	10
WorldView-1	DigitalGlobe USA 2007	494	0.45/–	17.6	pan 0.45–0.90	11
WorldView-2	DigitalGlobe USA 2009	770	0.46/1.84	16.4	pan 0.50–0.80 MS: 0.40–0.45, 0.45–0.51, 0.51–0.58, 0.58–0.625, 0.63–0.69, 0.705–0.745, 0.77–0.895, 0.86–1.04	11
GeoEye-1	GeoEye Inc. USA 2008	681	0.41/1.65	15.2	pan 0.45–0.80 MS: 0.45–0.51, 0.51–0.58, 0.655–0.69, 0.78–0.92	11
Cartosat-2, 2A, 2B	ISRO India 2007–2010	631	0.82/–	9.6	pan 0.5–0.75	10

With the exception of Resurs-DK1, all very high resolution optical satellites, as with most other optical satellites, are operated in a sun-synchronous orbit, usually collecting data in the descending orbit, with an equator crossing approximately between 9:30 and 11:00 o'clock in the morning. Because of priority access by a small number of large volume users, the available imaging capacity of high resolution optical satellites is limited.

5.2.2 IKONOS-2

IKONOS-2 was launched by the US company Space Imaging in 1999. Because of financial problems, Space Imaging was bought by OrbImage in 2005, and the name was changed to GeoEye Inc. The operational use of IKONOS images, after half a year of calibration and system set-up, was the start of high resolution optical space images in the civilian area.

IKONOS-2 has a GSD of 0.82 m for panchromatic and four times as much for the multispectral bands. This relationship of four between the resolution of panchromatic and multispectral is typical for space images. By pan-sharpening (see Section 7.6.2), the lower resolution colour information can be merged with the panchromatic channel to create a colour image with the panchromatic resolution. Under normal conditions, if the colour CCD line has the correct geometric reference to the panchromatic CCD line, no difference of the pan-sharpened image versus an original colour image with the same resolution can be detected.

When IKONOS started, the operating license from the US government required a resampling to 1.0 m. For this reason IKONOS images are usually distributed with 1.0 m GSD. Later, the limitation for the distribution of images outside the US government and to special users was reduced to 0.5 m, so IKONOS images with the full resolution can also be ordered today. Nevertheless, for an incidence angle of 25° the ground resolution is still 1.0 m (Figure 4.3, Equation 4.1), so no real loss of information was caused by this limitation. The former rule to hold back the images for 24 hours, for national US security reasons, is no longer in place. By the US Kyl-Bingaman Amendment to the Defence Appropriations Act in 1996, imagery over Israel is not distributed having a higher resolution than generally available; thus over Israel there is a restriction of 2 m GSD for US images.

As can be seen in Figure 5.4 the panchromatic band of IKONOS is extended to near infrared. This has advantages especially for imaging forest areas, where the dominant reflection is in the near-infrared range. The use of blue, green, red and near infrared dominates for very high resolution space images, allowing the generation of true colour and also false colour images. The spectral sensitivity of QuickBird corresponds to IKONOS, with both sensors using the same Kodak TDI arrays. As standard, both systems use TDI, usually with 13 steps.

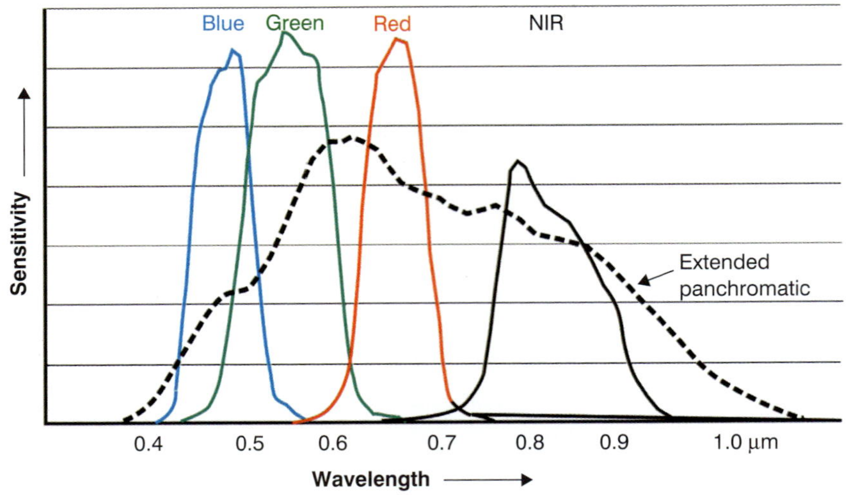

Figure 5.4 Spectral sensitivity of IKONOS and QuickBird bands.

If an area with a larger dimension than the swath of 11 km is required, IKONOS can image it first with the swath width, rotate the satellite in the across-orbit direction corresponding to the swath width, and then scan the remaining part against the movement in the orbit (Figure 5.5c). The imaging in the orbit direction is referred to as "reverse scan" because of the location of the used TDI array combination in the focal plane; the scan against the orbit direction is referred to as "forward scan". IKONOS-2 does not have a single linear array covering the whole swath; it has a combination of three TDI arrays (Figure 5.6). This TDI array works only in one direction, so for scanning in the opposite direction an additional combination of TDI arrays is required.

This is not the case for the colour band: because of the larger GSD, only CCD lines are used and they are independent of the scan direction. The panchromatic and the colour information can only be joined in the object plane. Due to the shift of the sensors in the focal plane, the sub-images correspond to different projection centres, related to different instants of exposure (Figure 5.7). According to theory, in this case a correct fit of the sub-images is only possible for one height level (see also Figure 4.22), but under operational conditions this is not a problem. For a misfit in the panchromatic image of one pixel, there must be a height difference of approximately 5 km from the reference height in the scene. For the combination of panchromatic and multispectral images, a misfit of one multispectral pixel appears in the case of a height difference of approximately 2 km from the reference height.

In practice, the sub-images fit together well; there is also a good fit of the multispectral information with the panchromatic information in pan-sharpened images. Only in the case of moving objects, such as cars on a main road, does the time difference of exposure cause a misfit of the colour information to the panchromatic

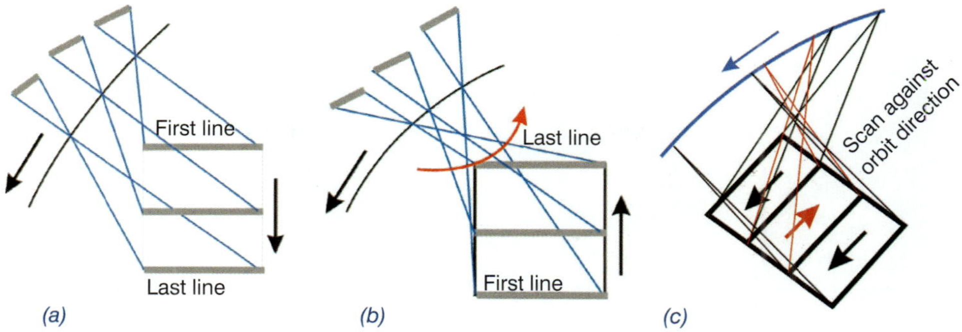

Figure 5.5 Scan modes for IKONOS: (a) scan in orbit direction (reverse); (b) scan against orbit direction (forward); and (c) combination.

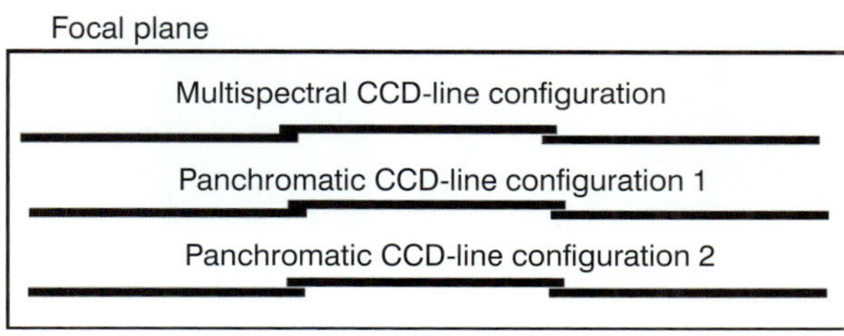

Figure 5.6 TDI array combination in the focal plane.

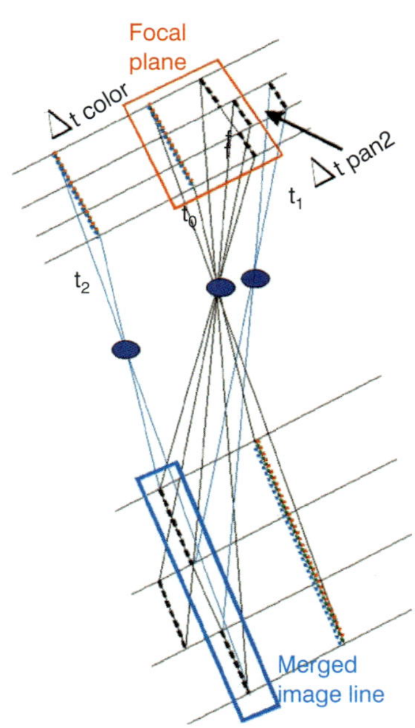

Figure 5.7 Sub-image connection for generating a pan-sharpened image.

(a) (b)

Figure 5.8 Shift of colour with respect to panchromatic information for moving cars: (a) IKONOS image; and (b) QuickBird image.

information; in reverse scan images, the colour follows the panchromatic information. The car speed can be estimated using the difference between both images (Figure 5.8). IKONOS scenes usually have good image quality and no loss of effective image resolution against the nominal has been detected by edge analysis (Figure 4.7). To enable high image quality and light sensitivity, the primary mirror of the optical system has a diameter of 71 cm.

IKONOS scenes are distributed as "IKONOS Geo", originally called "CARTERRA Geo". IKONOS Geo scenes are level 2A type images (Figure 5.10), projected onto a plane of constant height and related to a map projection. Level 1A type images, close to the original images, are not distributed with the inner orientation information. Initially, Space Imaging tried to sell value-added products such as orthoimages. Because of the high price, however, georeferencing solutions not using information about orbit data and inner orientation information were developed by universities and software companies.

As with all types of very high resolution space images, a specular reflection of the sunlight into the direction of the sensor causes an over-saturation of the CCD elements. In the case of sensors without TDI arrays, the "blooming" extends in all directions, enlarging the object to a blob; in addition, TDI sensors smear it in the TDI direction (Figure 5.9). Such effects can often be seen, especially in built-up areas, but they do not cause real

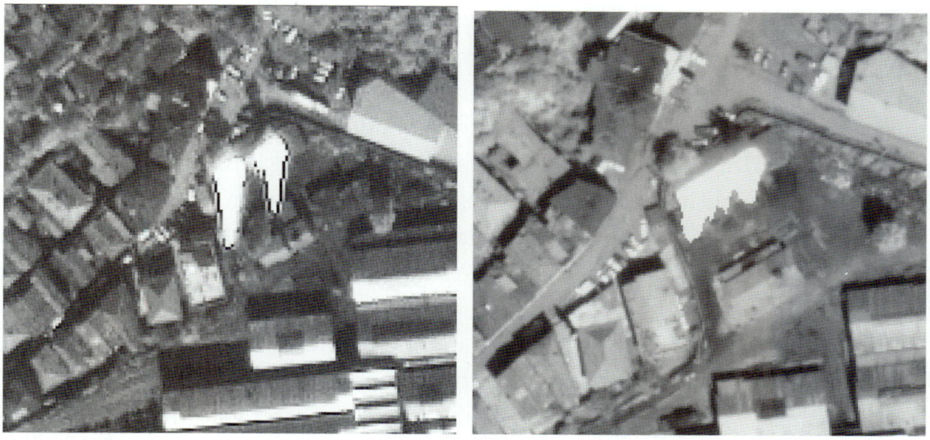

(a) (b)

Figure 5.9 Over-saturation (blooming) in high-resolution images: (a) IKONOS image; and (b) QuickBird image.

Table 5.3 Technical specifications of GeoEye products

Product	Geo	GeoProfes-sional	Precision	GeoStereo	Precision (stereo)
Accuracy (rmse) (m)	8	5	2	22	6
Ortho corrected	No	Yes	Yes	No	No
Remarks	Radiometrically corrected map-oriented image; standard product, no basic imagery available	Ground control points required	Limited to 18° incidence angle	Stereo pair	Ground control points required

problems. The dark line around the area of over-saturation in the IKONOS image in Figure 5.9 is caused by a contrast enhancement of the IKONOS scene. Such an enhancement is frequently observed in space images: although it improves the visibility of objects, it also changes the original grey values and is therefore not optimal for all applications.

5.2.3 QuickBird-2

QuickBird-2 was launched by DigitalGlobe in 2001 (DigitalGlobe was formerly EarthWatch before 2001). QuickBird was originally planned for a flying height that enabled 0.82 m GSD. With the US government's change in the allowance for distributing space images with 0.5 m instead of 1 m GSD, the flying height of QuickBird was reduced to approximately 450 km, to enable 0.61 m GSD for nadir view. At the time of the decision to change the flying elevation, it was too late to increase the sampling rate of the panchromatic sensor from 6300 lines/sec to 11 700 lines/sec, corresponding to 0.61 m line width in the flight direction. QuickBird compensates for this with an asynchronous imaging mode having a slow-down factor *B/A* of approximately 1.7 (Figure 4.5). From 2001, up until the launch of WorldView-1 by DigitalGlobe in 2007, QuickBird-2 delivered the highest resolution optical space images for civilian applications.

QuickBird-2 has the same TDI arrays and CCD lines as IKONOS-2, but instead of a combination of three TDI arrays to a homogenous virtual line, it has six of them, with 27 552 pixels leading to a swath width of 16.8 km for a view in nadir direction, even with the relatively small 0.61 m panchromatic GSD. This corresponds to a field of view twice as large as IKONOS-2. Furthermore, a multispectral CCD-line combination from Kodak, the same as for IKONOS, is used, resulting in the same spectral sensitivity (Figure 5.4).

QuickBird images are distributed as basic, standard and ortho ready standard (OR standard) imagery. Basic imageries are level 1A type homogenous merged images (Figure 5.10) of the six sub-images, corrected by the geometric sensor calibration as well as by the radiometric calibration. The orientation of such original images is supported by information about inner orientation, orbit and attitude data. Standard imageries are projected to the GTOPO30 height model, which is not very precise (see Section 7.4.1), so they are not too dissimilar to orthoimages. Of course, orthoimages

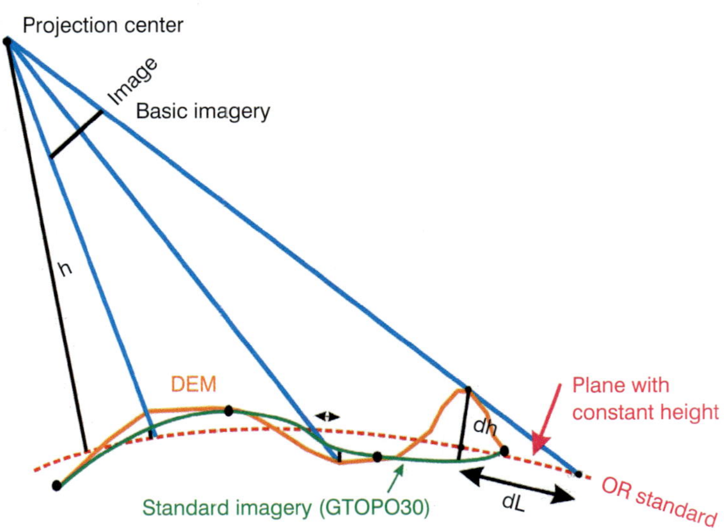

Figure 5.10 QuickBird image types.

can also be generated with standard imagery based on the height difference of a satisfactory height model from GTOPO30, but this is even more complex, as when using level 2A type images. Shortly after the start of production, by user request the OR standard was also offered; this is a level 2A type scene, projected onto a plane of constant height. For a standard georeferenced application, basic as well as OR standard imagery can be used.

5.2.4 OrbView-3

OrbView-3, from the US company GeoEye (formerly OrbImage), was launched in June 2003 and active until March 2007. This small satellite of 304 kg was significantly less expensive than IKONOS and QuickBird. The panchromatic sensor focal plane contains 16 sub-chips, with 505 CCD elements per sub-chip. The sub-chips are staggered in two rows (Figure 4.4a), resulting in 8032 active CCD elements for imaging. The staggered principle leads to 2 m projected pixel size in nadir view and, by 50% over-sampling, to 1 m GSD (Figure 4.4b), but the staggered imaging cannot produce the same image quality as scenes with 1 m projected pixel size. A direct comparison of IKONOS with OrbView-3 images of the same area shows clearer detail for IKONOS; for OrbView-3 scenes, approximately 30% less information content can be identified in the image. The multispectral sensor focal plane contains two sub-chips, with 1008 detectors per sub-chip, aligned in a row for each band (Figure 5.11), leading to a swath of 8 km in a nadir view direction.

Instead of TDI arrays, OrbView-3 has CCD lines, requiring a lower sampling rate for a satisfactory signal-to-noise ratio. The size of the panchromatic CCD elements is limited to 6 µm × 5.4 µm, and to 24 µm × 22 µm for the multispectral bands. The sampling rate of a maximum 5000 panchromatic lines per second or 2500 multispectral lines per second, leads to an asynchronous imaging mode (Figure 4.5) with a slow-down factor of at least 1.42 for the panchromatic mode. An asynchronous imaging

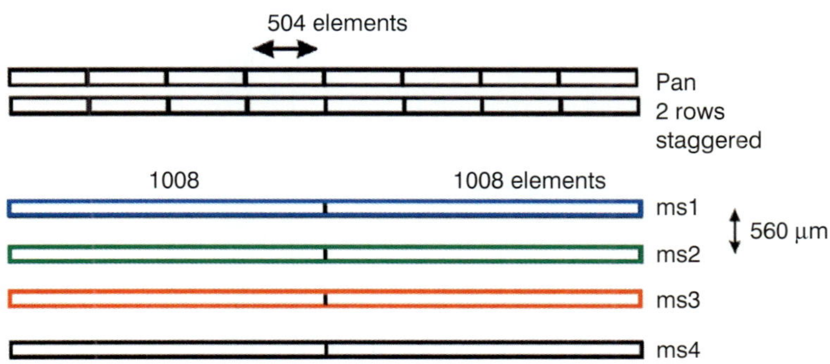

Figure 5.11 OrbView-3 focal plane arrangement.

mode reduces the imaging capacity, but in general this is limited to 10 minutes per orbit. In addition, only panchromatic or multispectral scenes can be taken at any one time; an acquisition of both together is not possible. As with IKONOS and QuickBird, OrbView-3 can also scan an area across the orbit. This is a typical configuration for stereo models, in which scanning occurs east to west, and later, from the same orbit, from west to east (or vice versa).

The extended panchromatic band is sensitive in the spectral range 0.45 to 0.90 μm, while the multispectral bands cover the wavelengths: blue from 0.45 to 0.52 μm, green from 0.52 to 0.60 μm, red from 0.63 to 0.69 μm and near infrared from 0.76 to 0.90 μm.

5.2.5 EROS B

ImageSat International, officially located in the Netherlands Antilles but controlled by Israel, originally announced that it would launch a series of six EROS B satellites, but finally this was cut back to one EROS B satellite in 2006 after a long delay, and will be followed by EROS C with a different specification. EROS B is a small satellite of 290 kg, limited to the extended panchromatic range of 0.5 to 0.9 μm wavelength, with 0.7 m GSD and a 7 km swath in a nadir view direction. The maximum scan rate is 2400 lines per second, requiring an asynchronous imaging mode with a slow-down factor of at least 4.2. The body pointing usually operates in the range of ±45° nadir angle. A 32 Gb recording capacity enables operation outside the range of the ground stations.

Without any radiometric correction, EROS B scenes may show a very strong striping, caused by staggered CCD lines. Figure 5.12(a) shows the grey value difference of neighbouring CCD lines over the whole scene width. This shows very clearly that EROS B is equipped with 8 × 2 CCD lines, and, especially, that the first and the fourth staggered CCD-line combinations have clear shifts in the grey value. Figure 5.12(b) shows the grey value difference between even and odd lines, averaged over some neighbouring line combinations, as a function of the grey value for four different locations in the scene. The strong dependency on the grey values is evident.

With a shift of the grey values in even and odd columns by half the difference of the average grey values of neighbouring columns, followed by a change of the grey values as a function of the difference of neighbouring columns depending on the grey values, averaged over some neighbouring columns (Figure 5.13b), the major effect of the striping can be eliminated and the objects can be seen more clearly. An inverse Fourier transform of the modified power spectrum (Figure 5.13c) leads to slightly better results.

5.2.6 KOMPSAT-2

The South Korean Space Agency KARI has a complete programme of Earth observation satellites, including some in development for the future. KOMPSAT-2, a high resolution optical satellite with 1 m GSD panchromatic and 4 m GSD multispectral

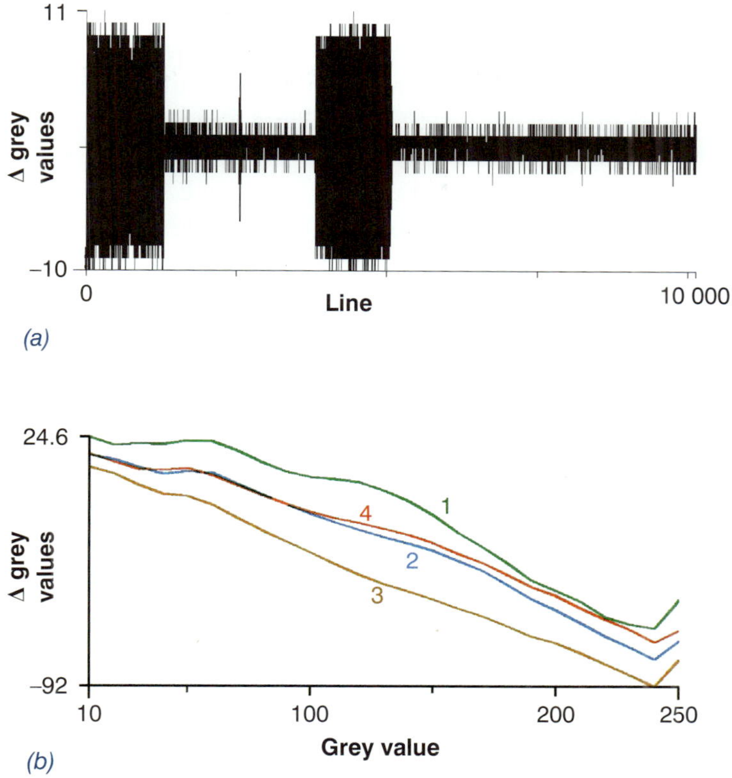

Figure 5.12 EROS B scenes: (a) grey-value difference of adjacent CCD lines; and (b) grey-value difference of CCD lines as a function of the grey value.

(a) *(b)* *(c)*

Figures 5.13 Portion of an EROS B image before and after de-striping: (a) original sub-image; (b) after de-striping as a function of grey values; and (c) using inverse Fourier transform of modified power spectrum.

resolution, was launched in 2006. Similar to the high resolution satellites mentioned above, the panchromatic range is extended to the near infrared with 0.50 to 0.90 μm wavelength, and the multispectral bands are blue (0.45–0.90 μm), green (0.52–0.60 μm), red (0.63–0.69 μm) and near infrared (0.76–0.90 μm). Panchromatic and multispectral imaging can be operated simultaneously, however. The 800 kg satellite was developed in a cooperation between KARI and the European company EADS Astrium.

Like all high resolution satellites, the imaging is made by body pointing of the whole satellite, usually with nadir angles of up to 45°. The high acquisition capacity reaches 20 minutes per orbit of uniform data strips, or 7500 scenes per day with 15 km × 15 km scene size. The sampling rate of up to 8000 lines/second does not require an asynchronous imaging mode and is one of the reasons for the very high collection rate. KARI prefers the sun-synchronous orbit in ascending mode, which differs from nearly all other satellites, but fits the shape of Korea better. An on-board image recording with a capacity of at least 1000 km image length and 15 km swath enables the use of the satellite outside the range of ground stations.

Stereo pairs can be generated across track. Four panchromatic TDI arrays, with up to 32 stages and a length of 5200 elements, are joined together to the virtual swath of 15 000 pixels. In the orbit direction, the distance of the TDI arrays corresponds to 308 m, so a misfit of 1 pixel appears at the joints of sub-images when there is an object height difference of 2.2 km from the reference height. This never occurs under normal conditions (see also Figure 4.22).

5.2.7 Resurs-DK1

The Russian Resurs-DK1, operated by Russia's Science Centre for Remote Sensing of Earth (NTs OMZ), does not directly belong to the group of high resolution optical satellites available for civilian purposes. In contradiction to the original announcement, the images are not internationally distributed. At 6.8 tons, Resurs-DK1 is a very heavy satellite in an elliptical orbit, with an altitude that varies between roughly 330 km and 585 km, leading to a variable GSD of approximately 0.8 m for the panchromatic band at a flying altitude of 350 km and 2 m up to 3 m for the three multispectral bands at the same height. The panchromatic band is sensitive in the wavelength range 0.58 μm to 0.8 μm, and the multispectral channels cover green (0.5–0.6 μm), red (0.6–0.7 μm) and near infrared (0.7–0.8 μm). The panchromatic band is based on a combination of 36 TDI arrays, each with 1024 CCD elements with up to 128 stages, leading to a combined line length of 36 000 elements. The orbit inclination of 70° is not sun synchronous and a body pointing mode is possible up to ±30° nadir angle.

5.2.8 IRS Cartosat-2

India has a very complete system of Earth observation satellites of different resolutions and specialisations. IRS Cartosat-2 was not the first very high resolution optical satellite from ISRO; the experimental satellite TES was launched in 2001, generating 1 m GSD for panchromatic images, using the asynchronous imaging mode. Cartosat-2 was launched in 2007, followed by the almost identical Cartosat-2A in 2008 and Cartosat 2B in 2010. Cartosat-2A is reserved for the Indian intelligence service, while Cartosat-2 is used for civilian applications. These satellites are limited to a panchromatic mode in the spectral range of 0.5 to 0.85 μm wavelength, with 0.82 m GSD and a swath of 9.6 km for nadir viewing. The 680 kg satellite usually operates with up to ±45° nadir angle. ISRO distributes the images via the company ANTRIX, who likes to be competitive with other high resolution satellite images within India, by offering imagery at prices that are lower by a factor of five to eight. The 64 Gb solid-state recorder can store 138 scenes of 9.6 km × 9.6 km, allowing imaging outside the range of the ground stations. Cartosat-2 has a CCD-line combination with 12 000 pixels. Even with the large diameter of 700 mm for the primary mirror, an asynchronous imaging mode is required.

5.2.9 WorldView-1

WorldView-1 from DigitalGlobe is a supplement to and successor of QuickBird-2. With 0.45 m GSD for 17.6 km swath, it belongs to a new class of powerful high resolution optical satellites, together with GeoEye-1. Not only has the ground resolution been improved, along with a larger swath, but also the agility and imaging capacity exceeds the capability of earlier launched satellites. A total area of up to 750 000 km² can be imaged per day. An area of 60 km × 110 km can be taken from the same path with four parallel strips, or a stereoscopic coverage of 30 km × 110 km with two parallel strips or strips up to 17.6 km × 246.4 km. The on-board solid-state recorder with a storage capacity of 2199 Gb enables imaging that is independent of the location of the ground station. Because of the US license, only images with 0.5 m GSD are distributed, but this corresponds to the projected pixel size for 25.8° incidence angle across the view direction or 18.4° in the view direction, so there is very often no loss of information versus the 0.45 m GSD in nadir view direction. The nadir angle range of ±45° can be exceeded by request.

This heavy satellite of 2500 kg is very agile, rotated by control moment gyros with a slew speed that enables pointing from one ground location to another, 300 km away, within 10 seconds; in comparison, QuickBird requires 62 seconds, and IKONOS and GeoEye-1 require 25 seconds (see also Figure 5.3). The slewing over a ground distance depends on the integrated rotation speed of the satellite and the flying height above ground. Because of its higher flying altitude, WorldView-2 has a faster slewing over the ground than WorldView-1, even though the rotation of WorldView-2 is slightly slower.

WorldView-1 is limited to the extended panchromatic band of 0.38 to 0.88 μm wavelength. It was the first satellite launched under the NextView contract of the US National GeoSpatial Intelligence Agency with a pre-order of images valued at US$500 million. The second NextView contract, of about the same amount, was made with GeoEye.

5.2.10 WorldView-2

DigitalGlobe's WorldView-2 was launched in 2009 and supplements WorldView-1. The ground resolution of 0.46 m for panchromatic is similar, but WorldView-2 has eight spectral bands (Table 5.4), with 1.84 m GSD for nadir view. WorldView-2 also has similar agility to WorldView-1, but has a flying height of 770 km instead of 494 km. The higher flying elevation has the advantage of a higher revisit frequency of 1.1 days at 1 m GSD, or less than 3.7 days at 20° off-nadir view, whereas the revisit is 1.7 days at 1 m GSD for WorldView-1. Both satellites are equipped with TDI arrays with up to 64 stages.

5.2.11 GeoEye-1

GeoEye-1, originally called OrbView-5 during the planning phase, was launched in September 2008 by the US company GeoEye. With 0.41 m panchromatic GSD and 1.65 m GSD for the multispectral range, it is currently the civilian space observation system with the highest ground resolution. Nevertheless, the US government limits the image distribution to 0.5 m GSD for all customers not explicitly granted a waiver by the government. The construction of its successor, GeoEye-2 (to be launched in 2012), has already started, and this satellite will further improve the ground resolution to 25 cm.

The panchromatic range extends from 0.45 to 0.80 μm wavelength; the multispectral bands are blue (0.45–0.80 μm), green (0.51–0.58 μm), red (0.655–0.69 μm) and near infrared (0.78–0.92 μm). The light sensitivity of the 1.1 m diameter of the primary mirror is supported by TDI arrays with up to 64 stages, covering in total a swath of 15.2 km in nadir view direction.

The 1955 kg satellite is equipped with high performance reaction wheels, allowing a slewing at approximately the same speed as IKONOS (Figure 5.3). The imaging capacity reaches up to 350 000 km^2 per day for panchromatic and multispectral simultaneously, or 700 000 km^2 per day for panchromatic only. Stereo scenes of up to 224 km × 28 km or mono scenes of up to 300 km × 50 km can be taken from the same track. From the 681 km high orbit, GeoEye-1's revisit capability with a nadir angle of 28°, corresponding to 0.5 m GSD, is 2.8 days; for 35° nadir angle, it is 2.1 days. With a 1000 Gb recorder it can operate outside the range of the ground stations.

Table 5.4 Spectral sensitivity (μm) of WorldView-2 bands

	Pan	Coastal	Blue	Green	Yellow	Red	Red edge	NIR1	NIR2
From	0.45	0.40	0.45	0.51	0.585	0.63	0.705	0.77	0.86
To	0.80	0.45	0.51	0.58	0.625	0.69	0.745	0.895	1.04

The construction and operation of expensive Earth observation satellites such as GeoEye and WorldView is only possible with guaranteed orders from governmental agencies such as the US National GeoSpatial Intelligence Agency (NGA), bringing in 55% of the revenue. Therefore the NGA is given priority for image acquisition, which means that in some areas it may take a while to get images. As in the case of WorldView, GeoEye-1 is a fully redundant system in order to guarantee the expected lifetime of at least seven years. It cost approximately US$502 million to bring the satellite into service.

5.2.12 Announced systems

The number of very high resolution optical satellites will grow in the future and these systems are discussed in Chapter 8. Of course, some details may change – the launch dates may be postponed, some systems may be cancelled or fail, or additional systems may be added.

References

Dial, G. (GeoEye) 2008. GeoEye: Remote sensing & production services. Civil Commercial Imagery Workshop, Fairfax, VA, USA, http://calval.cr.usgs.gov/JACIE_files/JACIE08/2008_JACIE_DVD/Wednesday_PM/Dial_GeoEye.pdf (accessed February 2011).

Hellermann, R. (ImageSat) 2008. EROS B – Operational experience, second year of operation. Civil Commercial Imagery Workshop, Fairfax, VA, USA.

McGill, M. (DigitalGlobe), 2005. A US perspective of Phase 2 of high resolution satellite remote sensing. Eurimage Meeting, Rome, Italy.

Smiley, B. (DigitalGlobe), 2008. The absolute and relative geolocation accuracy of QB02 and WV01. Civil Commercial Imagery Workshop, Fairfax, VA, USA, http://www.asprs.org/publications/proceedings/baltimore09/0092.pdf (accessed February 2011).

6 Calibration, Sensor Models and Orientation

Karsten Jacobsen

6.1 General information

Data acquisition based on satellite images requires a correct referencing of the identified objects. This is only possible if the relationship between object coordinates and image coordinates is known. With the exception of some approximate methods of georeferencing, using general polynomials or simplified mathematical models for the relationship of image to object coordinates, the inner orientation must be known, at least indirectly using the field of view. Inner orientation defines the relationship between the projection centre and the image position and requires knowledge of the principal distance, principal point and lens distortion, found from calibration of the sensor. In the case of CCD-line images, the image position is limited to a line of pixels. The inner orientation, together with the line of pixels, determines the bundle of rays from the image through the projection centre to the ground positions. For some sensors, for example IKONOS, no information about the inner orientation is distributed, requiring the reconstruction of the bundle of rays based on the object position and the corresponding projection centre (Figure 6.1).

The reconstruction of the bundle of rays with the projection centre as origin is based on the mathematical model of perspective so that the projection centre **O**, the image point **a** and the object point **A** all fall on a straight line; the line of detectors must also be a straight line. The checking of the alignment of the image pixels is part of the process of geometric calibration (Section 6.2), while the 3D orientation of the bundle of rays in the object space belongs to the sensor orientation, required for the correct determination of the exterior orientation. The relationship of one line of points imaged on the ground to the neighbouring lines in the case of CCD-line images is part of the exterior orientation; this is supported by the direct sensor orientation of Earth observation satellites, based on a global navigation system (such as GPS), star sensors and gyros for the attitude determination as shown in Figure 6.2. With gyros, the image rotations can be determined with high frequency, but the long-term accuracy is not good and so

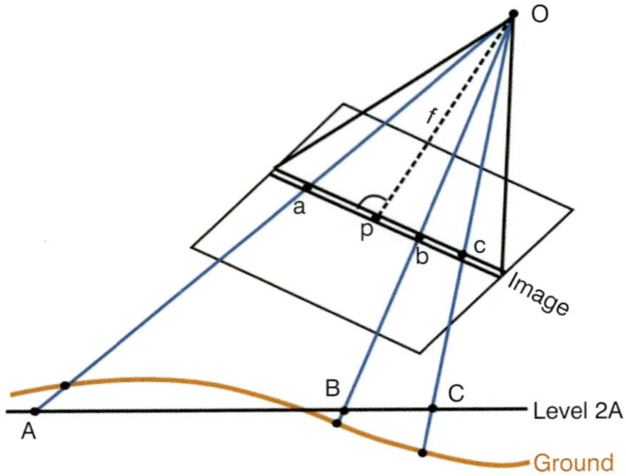

Figure 6.1 Inner orientation of a line sensor and relationship to a level 2A type image (O = projection centre; p = principal point; f = focal length).

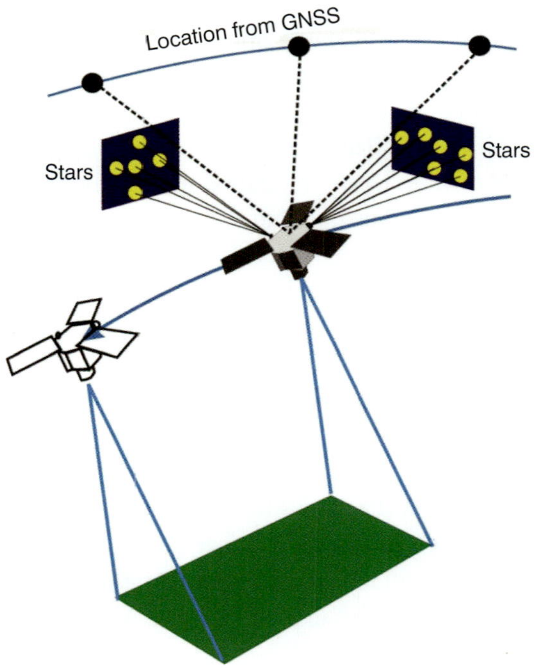

Figure 6.2 Direct sensor orientation by a global navigation system, star sensors and gyros.

have to be supported by stellar cameras giving the absolute angular orientation with lower frequency. The direct sensor orientation also has to be checked, leading to a system calibration.

The geometric calibration as well as the sensor orientation uses the mathematical concept of collinearity, which is based on a sensor with a distortion-free optical system, a flat focal plane and the ray from the object point through the projection centre to the image being a straight line. However, the path of light through the atmosphere is affected by refraction and therefore may not be a straight line. For space images the refraction is not a major factor, because in the most important part of the ray, close to the sensor, there is no atmosphere. Equation (6.1) gives the refraction correction for image coordinates of nadir images based on ARDC 1959 standard atmosphere [Jacobsen, 1986].

$$dr = \left(0.113.\frac{(P_1 - P_2).Z_g}{(Z_f - Z_g)} - \left(\frac{Z_f.2410}{(Z_f^2 - Z_f).6+250} - \frac{Z_g.2410}{(Z_g^2 - Z_g).6+250}\right).\left(r + \frac{r^3}{f^2}\right)\right).10^{-6} \quad (6.1)$$

where

dr is the displacement on the image due to refraction,
Z_f is flying height above mean sea level (km),
Z_g is mean terrain height above mean sea level (km),
f is focal length (mm),
r is radial distance in nadir image (mm),
P_1 is air pressure (mb) in terrain height,
P_2 is air pressure (mb) at flying height $P = e^{(6.94-Z\cdot 0.125)}$ (mb).

The refraction correction given by this equation was developed for space photographs taken in the nadir direction. For an inclined view this has to be transformed into an angular influence or the influence on the ground position. Figure 6.3 shows the influence of the refraction on the ground position at mean sea level as a function of the nadir angle. For higher terrain elevation, the refraction influence is smaller.

The influence of the refraction is limited: for example, the field of view (FOV) for Resurs-DK1 reaches only 4.6°, but even with such a FOV, the change in the influence of refraction within the FOV is negligible. In the case of a very large slow-down factor

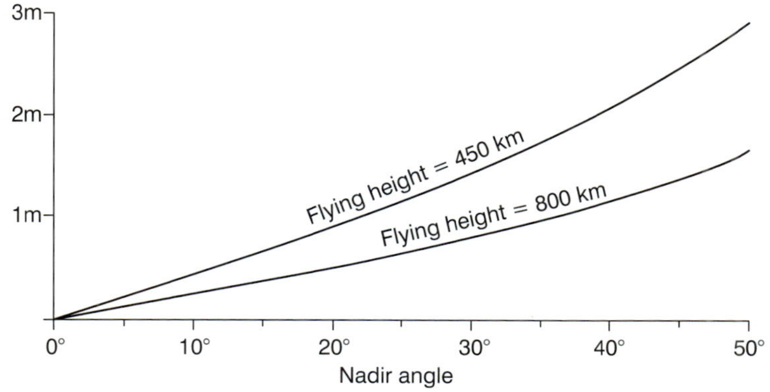

Figure 6.3 Influence of refraction on ground position at mean sea level as a function of nadir angle.

for an asynchronous mode, the influence may become slightly larger, but is usually compensated for by georeferencing with ground control points (GCPs). The influence of refraction need only be taken into account for georeferencing without control points, which is possible in some cases with IKONOS-2, GeoEye-1 and WorldView, based on the very good absolute scene orientation by direct sensor orientation. However, it is always included in the rational polynomial coefficients (RPCs), which are discussed in Section 6.3.2.2.

6.2 Calibration

6.2.1 Radiometric calibration and image quality

The grey values in an image must correspond to the intensity of reflection from the corresponding ground element in the specified spectral range. Every pixel of a CCD line or the average of the used CCD elements of a TDI array has a slightly different sensitivity. For very precise radiometric calibration, any individual CCD element or row of a TDI array has to be investigated. This can be done before launch, but the sensitivity degrades over time, requiring post-launch calibration. In the case of post-launch, the influence of the atmosphere has to be taken into account. Larger differences may occur between the different CCD lines or TDI elements in the focal plane. Staggered lines (Figure 4.4a) in particular may show this, visible as a striping effect, and may cause a negative influence on the object identification and automatic image matching. The striping effect can be determined and corrected within a scene by a relative radiometric change.

Figure 6.4(a) shows an original ASTER sub-scene with obvious striping. For ASTER images striping is usually only a constant shift of the grey values. In this example, the average grey value difference of even and odd rows is 12.2 grey values. By a constant shift of the grey values in the odd rows by 6.1 grey values and in the case of the even rows by −6.1, the striping completely disappears (Figure 6.4b). As shown in Section 5.2.5, the striping is more complex for EROS B scenes and requires a more local correction as a function of the individual grey values.

In addition to a pre-flight radiometric calibration, a complete radiometric sensor calibration can be made with uniform test fields. A selection of 36 uniform test fields, distributed all over the world, is described by Chander (2008). Only beach areas, dry lake beds, salt flats and desert sand areas are included, because snow fields are dependent on atmospheric conditions, vegetation areas are changing and, for water surfaces, the bi-directional reflectance distribution function (BRDF) varies strongly. Based on such uniform test fields, the radiometric sensor response can be determined and respected for a correction of the grey values. Schroeder *et al.* (1999) describe the radiometric test field calibration in detail. Such a post-flight radiometric calibration is important for the band-to-band calibration, to avoid colour shifts, and for the absolute calibration for some spectral analysis applications. For very high resolution space

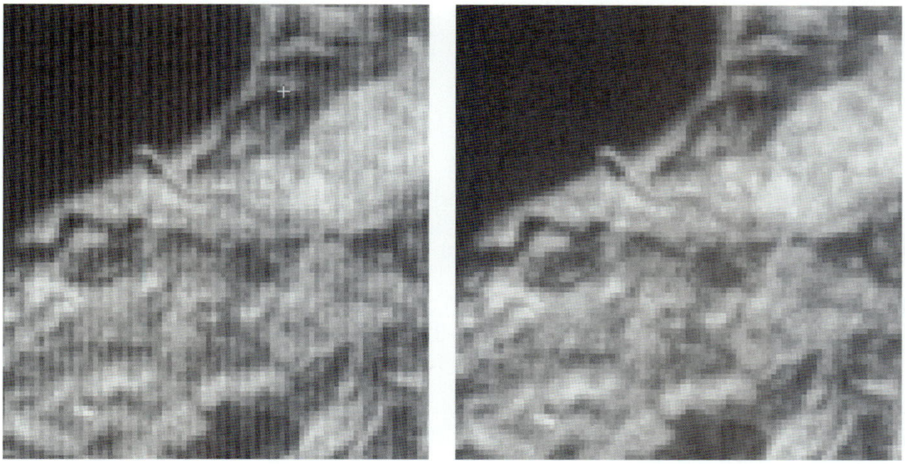

(a) (b)

Figure 6.4 An ASTER scene before and after de-striping: (a) original with striping; and (b) after de-striping.

images, the relative calibration is more important in order to avoid grey value variations in uniform areas, leading to striping and disturbing the object interpretation. Nevertheless, the strong image compression necessary for downloading images from the satellites may cause some low noise in uniform areas (see Section 3.6.4). Today such noise is smaller than when high resolution optical space images were first obtained.

For sensors with lower resolution, for example Landsat, primarily used for classification purposes, the radiometric calibration is more important. Here calibration devices, such as reference lamps or solar calibrators, are often available in the satellites. Even stars can be used for radiometric calibration [Tansock *et al.*, 2004] or the reflection over Antarctica snow fields.

Some random noise cannot be avoided, even with standard radiometric calibration, as shown in Figure 6.5. The random noise is of a low level of 2–5 grey values. Depending on the light conditions, it may vary as is obvious for both IKONOS sub-areas: the area IKONOS (1) has been imaged with higher sun elevation, while IKONOS (2) has been taken with low sun elevation. KOMPSAT-1 shows striping with a variation of up to five grey values, but in land areas it was not so disturbing as in the case of ASTER or EROS B.

Optionally, in some satellites, the grey value distribution can be adjusted, using dynamic range adjustment (DRA) or gain control, to a more uniform distribution over the usual 11-bit range of grey values. This may be good for manual image interpretation, but it destroys the absolute radiometric accuracy. For images optimised by DRA only, a relative grey value adjustment is important. An over-saturation (blooming), as shown in Figure 5.9, has nothing to do with calibration.

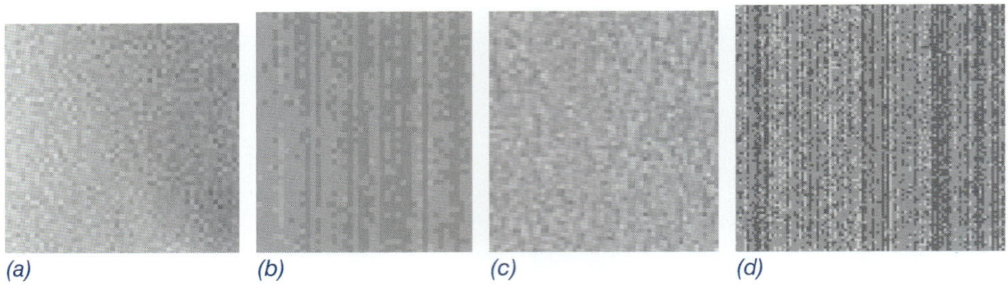

Figure 6.5 Random noise of space images in uniform areas (highly magnified): (a) IKONOS (1); (b) IKONOS (2); (c) OrbView-3; and (d) KOMPSAT-1.

6.2.2 Geometric calibration

With very few exceptions, high resolution optical satellites use combinations of CCD lines or combinations of TDI arrays. The centre line of the TDI array corresponds to the CCD-line location – no other geometric difference exists between CCD line and TDI array. The pixels in the CCD lines or the centre line of TDI arrays must be aligned and must have an equal spacing. The linearity can be investigated in the laboratory before launch, but due to the launch shock, dry-out under vacuum conditions and thermal influences, a deformation of the CCD lines may occur after launch, requiring a post-launch calibration for high accuracy requirements.

In Figure 6.6, the laboratory calibration of the panchromatic CCD line of the MOMS sensor shows a deviation of up to 2.2 pixels from the instrument axis. In relation to the CCD line itself, the discrepancies from a tilted straight line show the deformation of the CCD line: this is limited to 1 pixel across and 0.6 pixels in the CCD line. For the other six MOMS CCD lines having 6000 pixels, the deviation from a straight line was limited to a maximum of 0.5 pixels.

The geometry of the CCD lines has to be validated under flight conditions over a test field with ground control points. The SPOT satellites have been calibrated in-flight since the launch of SPOT-1 in 1986. In the case of SPOT, the mirror has a dominating influence as an additional source of deviation. The influence of the CCD-line linearity is usually below 0.2 pixels [Valorge *et al.*, 2004]. For OrbView-3 [OrbImage, 2004] discrepancies of the CCD line from a straight line of up to 8 pixels are reported. The size of the discrepancy is entirely unimportant provided it is determined by calibration and corrected.

With the exception of some micro satellites, all high resolution optical satellites have combinations of shorter CCD lines or TDI arrays as the base for the final virtual image line; in most cases they also use combinations of CCD lines in the focal plane (Figure 6.7). The combination of CCD lines forming a long virtual CCD line has to be calibrated to permit a transformation of the sub-images to a homogenous image without deformation within a line. As before, this can be calibrated in advance, but the

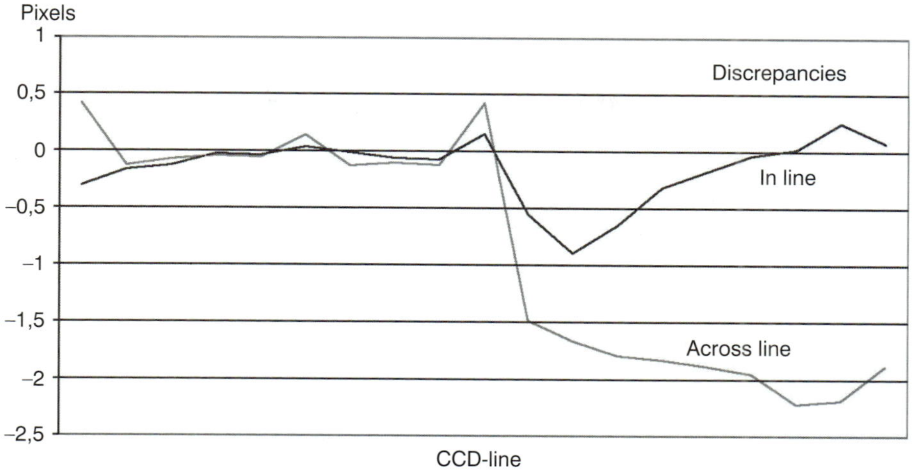

Figure 6.6 Laboratory calibration of a MOMS CCD line with 8650 pixels. Discrepancies in pixels can be seen in the 'in line' (black) and 'across line' (gray) directions.

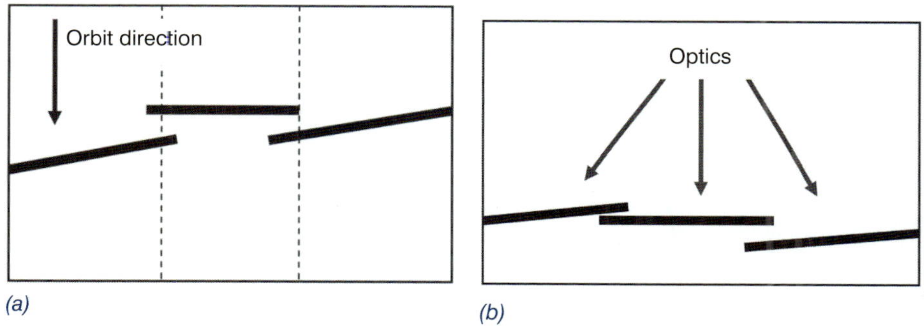

Figure 6.7 Location of individual CCD lines, to be determined by calibration: (a) in focal plane; and (b) perpendicular to optical axis.

calibration has to be verified under flight conditions and, if necessary, improved. In addition, the geometric relationship of the different CCD-line combinations has to be validated to ensure good band-to-band registration.

As was seen in Figures 6.6 and 6.7, a different location of one of the CCD lines or TDI arrays in the orbit direction corresponds to a different instant of exposure in relation to the virtual CCD line. If the object space is not very mountainous, this does not cause geometric problems of the merged sub-images (Figure 4.22). Nevertheless, the time delay of the different CCD lines has to be determined, along with the other possible geometric deviations and scale differences from a homogenous straight line of pixels with equal pixel size.

As can be seen in Figure 6.7, the following geometric differences may occur:

- the CCD lines may not be located in the focal plane, corresponding to a scale error;
- the CCD lines may not be parallel to the focal plane, corresponding to a slightly varying scale;
- the CCD lines may be rotated in the focal plane;
- the CCD lines may have a different instant of exposure, corresponding to a shift in orbit direction.

These geometric uncertainties have to be calibrated and validated. The sub-images of neighbouring CCD lines slightly overlap, thus allowing the shifts between the sub-images in line and sample direction to be determined. For the other unknowns, information from the object space is required.

Post-launch calibration is possible based only on ground control points, but it may be supported and improved by multiple coverage of the same area with different view directions, and also from different orbits, using rotation about the orbit direction. Some unknowns of a system calibration, including the direct sensor orientation, can only be determined with multiple coverage of the test area viewed from different directions, or using a real 3D test field with large variation of the control point height. As illustrated by the example of an IRS-1C calibration shown in Figures 6.8 and 6.9 [Jacobsen, 1997],

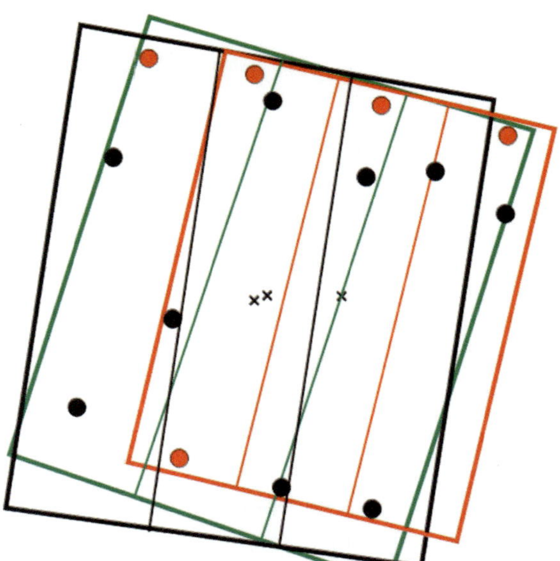

Figure 6.8 Footprints of three IRS-1C scenes used for calibration with images from three orbits (red circles show theoretically acceptable control points; dark circles are the control points used).

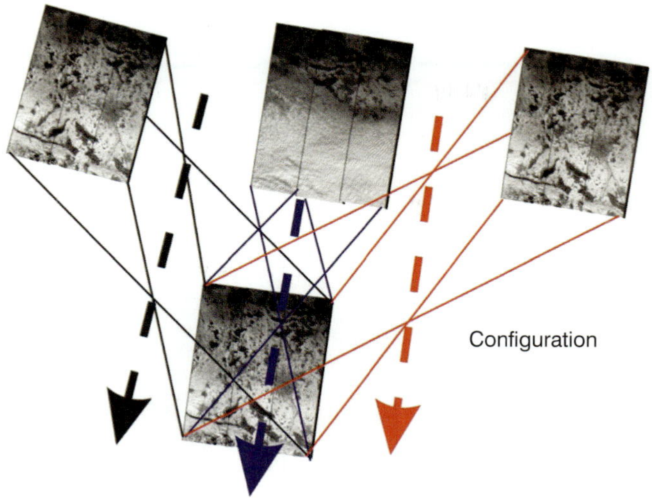

Figure 6.9 3D configuration of three IRS-1C scenes used for calibration.

with overlapping coverage of the sub-scenes and multiple complete scenes, a calibration of the CCD-line arrangement is possible using just a few ground control points [see also Srinivasan *et al.*, 2008].

The calibration and validation of the imaging system can be performed in two steps: the sub-scenes can be transformed all at once by means of tie points located in the overlapping area of the sub-scenes (Figure 6.10), followed by an adjustment of all scenes together with additional unknowns for the calibration and validation of the CCD-line configuration.

The adjustment of all scenes together of the IRS-1C data set is supported by tie points between the different scenes. Also the points tying the sub-scenes together have been used as tie points for the different scenes to stabilise the adjustment. The view from neighbouring orbits caused a convergent overlapping of the scenes as shown in Figure 6.8, leading to an improvement of the block configuration, so in theory, with just five control points (shown as red circles in the figure), the determination of all unknowns would be possible. In reality, the existing dark control points in Figure 6.8 have been used. As seen in Figure 6.9, the centre scene was partially covered by clouds, but this did not limit the adjustment. The results are shown in Figure 6.11.

The additional unknowns for the geometric relationship of the CCD lines included an individual rotation and an individual scaling of the sub-scenes, together with an affine parameter for the whole scene. The shift in line and sample direction was determined in advance in the image space. Based on the relationship of the CCD lines to each other, the sub-scenes can be transformed to a homogenous virtual scene, which may be also corrected by individual CCD-line deformation. Such virtual

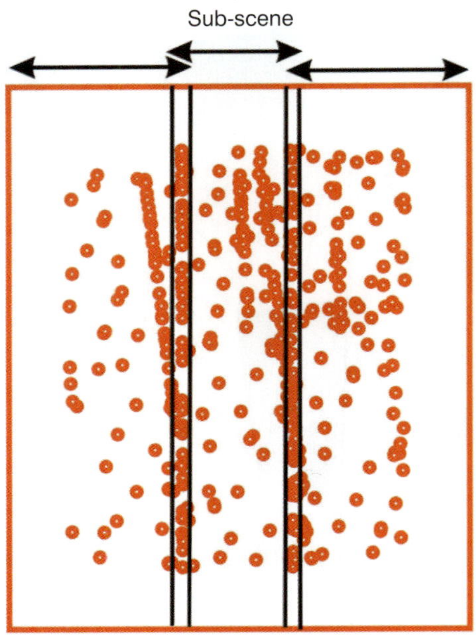

Figure 6.10 Point configuration used in one IRS-1C scene. Points located in overlapping parts of neighbouring sub-scenes are internal tie points, which are also used as tie points for other combined scenes.

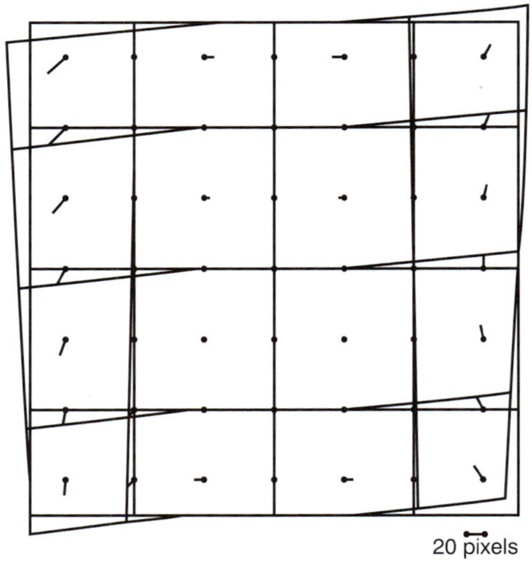

Figure 6.11 Geometry of joined sub-scenes as a result of self-calibration, showing discrepancies around the edges of the block.

scenes are distributed as level 1A type or basic imagery. A determination of the focal length of space images having a very small field of view is only possible with positions of the projection centres determined by a global navigation satellite system (GNSS), such as GPS.

Normally the user of space images will not receive the original sub-scenes. One exception is ALOS PRISM data, where original sub-images can be ordered. PRISM is a system of three cameras, viewing 24° forwards, to the nadir and 24° backwards. The nadir-looking camera has six CCD lines, while the inclined viewing cameras have eight. Neighbouring sub-images overlap by 32 pixels, allowing the use of tie points. The general handling of the adjustment of the CCD lines to each other also has to be included in a common adjustment, which is supported by the configuration of the three cameras [Kocaman and Gruen, 2007]. Sub-pixel accuracy can be achieved from processing such systems.

The generation of colour or pan-sharpened images requires knowledge of the geometric relationship between the different bands, or the so-called "band-to-band calibration". This relative geometric relationship can be determined based only on cross-correlation without control points. Instead of the collinearity equation used for absolute validation and calibration, the relative relationship can be based on coplanarity conditions such as a standard relative orientation [Srinivasan *et al.*, 2008].

For reaching a high accuracy level of absolute positioning, a complete system calibration has to be done, including the inner sensor geometry and the relationship of all positioning systems together. All high resolution optical Earth observation satellites are equipped with a global navigation satellite system (such as GPS), star sensors and gyros. Star sensors lead to an attitude determination with lower frequency. Higher-frequency rotations have to be interpolated based on gyros from the base of the star sensor information, which may be interrupted by viewing into the direction of the sun. Thermal radiation of the sun complicates the system geometry. A complete system calibration has to use test fields with a sufficient number of ground control points and must be done under different pitch and roll angles. The initialisation phase after satellite launch may take up to six months. To maintain a high accuracy of georeferencing without control points, the system calibration has to be validated and improved from time to time. The validation can be made over the same test field, enabling an automation of the whole process, including automatic control point identification [Valorge *et al.*, 2004].

Figure 6.12 shows the steps of geometric sensor modelling, adapted from Seo *et al.* (2008). Not all these steps will be done separately during calibration and validation. Here a common adjustment of most of the steps has advantages, but after determination of optimal calibration parameters, the sensor modelling will follow this sequence. Weser *et al.* (2008) give more information on satellite coordinate systems.

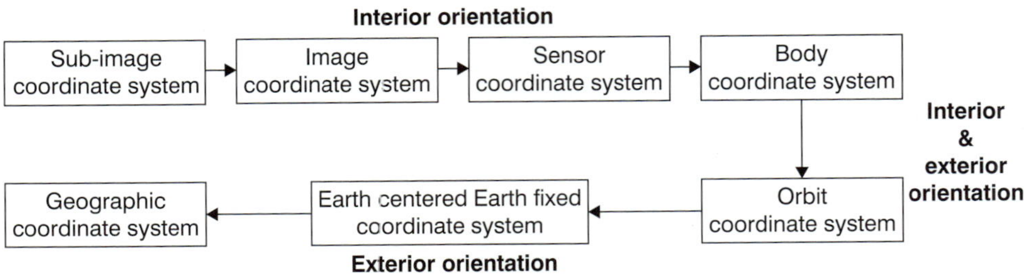

Figure 6.12 Sequential steps of sensor modelling (adapted from Seo *et al.*, 2008).

Some of the large very high resolution optical satellites reach very accurate absolute geopositioning without the use of ground control points. Absolute geopositioning depends on the nadir angle. Like the projected pixel size (Figure 4.2), the standard deviation of georeferencing in the view direction is enlarged by the factor $1/\cos^2 i$ (where i is the incidence angle), and in the across-track direction it is enlarged by the factor $1/\cos i$. For a better comparison, the absolute geolocation is reduced in the nadir view direction [Smiley (DigitalGlobe), 2008] and has to be enlarged for an inclined view direction. Also for a better comparison, in this chapter the accuracy is transformed to the standard deviation of the ground coordinates, because in the literature a confusing mixture of standard deviation and circular error with 90% probability (CE90) is used (as discussed in Section 1.2).

Dial and Grodecki (GeoEye) (2003) give the standard deviation of the absolute georeferencing accuracy of IKONOS without control points to approximately $\sigma_X = \sigma_Y =$ 4 m, which corresponds to the experience of the author. Smiley (DigitalGlobe) (2008) talks about an absolute standard deviation $\sigma_X = \sigma_Y$ for QuickBird of approximately 9 m; this can also be confirmed by the author. For OrbView-3 [OrbImage, 2004] the absolute standard deviation of X and Y will be 8 m. WorldView-1 [Smiley (DigitalGlobe), 2008] has reached a standard deviation of X and Y of 2 m; this corresponds approximately to the specification for GeoEye-1. This very high accuracy, without the use of control points, is related to WGS 84 in the International Terrestrial Reference Frame (ITRF) (see Section 1.2). In several countries the relationship between the local national datum and ITRF is not known to the same level of accuracy as the absolute georeferencing without control points.

Other sensors do not achieve this accuracy level. Srinivasan *et al.* (2008) specifies the initial absolute georeferencing for Cartosat-2 without in-flight calibration with a range of 1.1 to 1.3 km. After in-flight calibration it was improved to approximately 30 m. Cartosat-1 is still at the level of 200 m. For KOMPSAT-2, Seo *et al.* (2008) talk about an absolute positioning of approximately 35 m. If ground control points are available, an improvement to sub-pixel level is possible. These results are summarised in Table 6.1.

Table 6.1 Summary of georeferencing accuracy of selected sensors

Sensor	$\sigma_x = \sigma_y$ (m)	Reference
IKONOS	4	Dial and Grodecki, 2003
QuickBird	9	Smiley, 2008
OrbView-3	3	Orbimage, 2004
WorldView-1	2	Smiley, 2008
WorldView-2	3	DigitalGlobe
GeoEye-1	2	GeoEye Inc.
Cartosat-2 – no in-flight calibration	1100–1300	Srinivasan *et al.*, 2008
Cartosat-2 – with in-flight calibration	30	
Cartosat-1	200	
KOMPSAT-2	35	Seo *et al.*, 2008

6.3 Sensor models and orientation

6.3.1 Images with constant view direction across orbit

The first high resolution optical satellites did not change the view direction in relation to the orbit during imaging. This changed with agile satellites, which dominate the market today, and even with SPOT-5 the mirror is constantly rotated to compensate for the Earth rotation effect on the scene being imaged.

The sensor orientation, or exterior orientation, determines the relationship between the object and the image coordinates. The first operational optical space images used for civilian applications with accurate georeferencing came from SPOT-1 in 1986. The georeferencing of Landsat images, available before SPOT-1, was usually done approximately; this was sufficient for the ground resolution of Landsat imaging and due to the fact that Landsat views only in the nadir direction. SPOT and similar systems operate during imaging with a fixed view direction in relation to the orbit. In most cases the view direction can be changed using a rotating mirror; in some cases, such as IRS-1C/1D, a rotating camera, across the orbit, is used.

The first step of the transformation from the image to the ground coordinates is the inner orientation. In no case is the inner orientation published by the satellite operator, so if this is needed, it has to be reconstructed from the available information. The metadata or header data of SPOT and similar systems include information about the orbit location in the Earth-centred Earth-fixed coordinate system (ECEF – related to the Earth, rotating in the inertial space), together with attitude information. The orientation included in the satellite scene metadata based on a global navigation satellite system, star sensors and partially gyros is called "direct sensor orientation". This has to be interpolated using a time stamp to give the actual pixel position in the image. Different solutions for the interpolation of the satellite position and sensor attitude are in use; these range from interpolation in an inertial ellipse optimally fitted to the metadata information, transformed by rotation to a relative inertial coordinate system, to interpolation using a local best-fitting inertial circle, polynomials or piecewise polynomials, or cubic splines.

It is also possible to reconstruct the inertial orbit ellipse based on the generally known satellite orbit, together with the view direction and a ground control point. The time-dependent satellite position in the inertial orbit has to be related to the ECEF coordinate system, which changes by the Earth's rotation as a function of time. From the geocentric ECEF coordinate system, a transformation and a datum transfer to the national net coordinate system is required for the relation to the ground control points (Figure 6.13). Often the datum of the national net is not known very well. Some rough datum information is available on the Internet [NGA, 2011].

Perspective geometry exists in the plane defined by the CCD line and the perspective centre. It does not matter whether the image line is physically a homogenous CCD line or if it is composed of a group of shorter lines and treated as a virtual homogenous line. In the orbit direction the exterior orientation is different for every image line of the scene. It is not possible to determine the three rotations and the X, Y, Z coordinates of

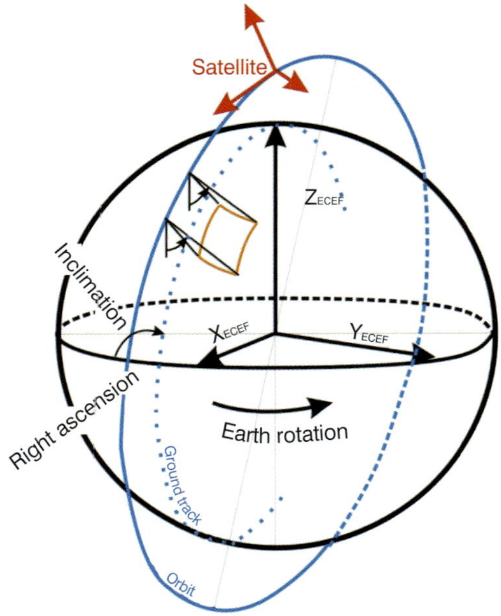

Figure 6.13 The relationship between space images with constant view angle across orbit, and Earth-centred and Earth-fixed coordinate systems.

the projection centre (the six elements of exterior orientation) for any line independently just based on image and ground control points for the whole scene, because the unknowns of all lines are strongly correlated to each other. The projection centre must be located on the orbit, for which position as well as attitude information is available, with spacing defined by the satellite operator. This orientation information has to be transformed to the ground coordinate system used as reference and improved by means of ground control points. In general the orbit is fixed in the inertial space and the Earth is rotating within, but the relationship between inertial space and the ECEF coordinate system is known.

With precise orientation information, the geometric accuracy of optical space images can be of the order of 1 GSD and better. Several, slightly different, software solutions are available for the orientation of the optical space images close to the originals – the level 1A-type or basic imagery. Because scene geometry is usually stable, in most cases a common change of the exterior orientation of all lines of a scene is sufficient. The small field of view of the space images and the CCD-line geometry makes it nearly impossible to separate the influence of roll and pitch from the projection centre location X_0 and Y_0. As can be seen in Figure 6.14, a change in roll and pitch causes only a shift of the footprint, leading to strong correlation between roll and pitch and projection centre coordinates X_0 and Y_0 if the ground control points are located approximately at the same elevation. The correlation coefficients often exceed 0.9995.

The correlation problem can be solved without loss of accuracy by selecting the unknowns to be determined for a scene. Jacobsen (1994) suggests determining only four orientation unknowns for a scene: these could be three rotations and the projection centre height Z_0, or the three projection centre coordinates and yaw. In addition, an affine parameter for correcting the flying height or speed and an angular affine

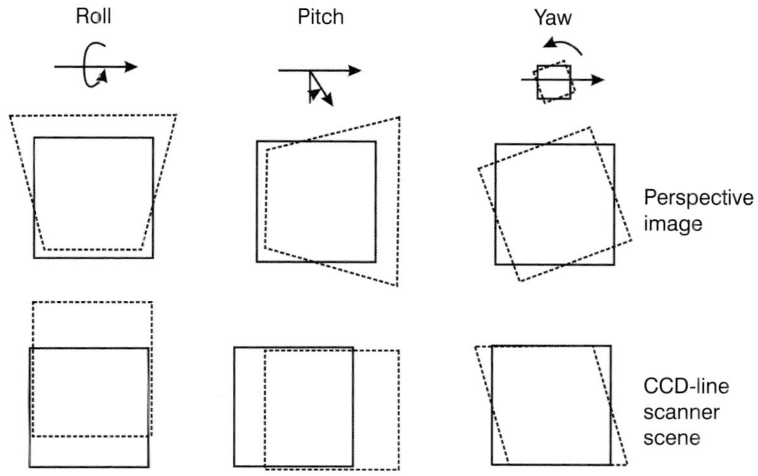

Figure 6.14 Effect of changing attitude data of image footprints.

parameter for correcting a yaw are required, so finally six orientation unknowns can be used, but note that these are not the classical photogrammetric unknowns.

There have been a number of formulations of the sensor model starting from the collinearity equations. For a single CCD line, the collinearity equations apply with a time factor introduced as given in the equation

$$\begin{bmatrix} 0 \\ y - y_o \\ -f \end{bmatrix} = \lambda \, . \, M_t \begin{bmatrix} X - X_{0_t} \\ Y - Y_{0_t} \\ Z - Z_{0_t} \end{bmatrix} \tag{6.2}$$

where t is the acquisition time of a CCD line; $0, y, -f$ are the image coordinates; X, Y, Z are the ground coordinates of the object point; $X_{0_t}, Y_{0_t}, Z_{0_t}$ are the object coordinates of the CCD-line perspective centre; M_t is a 3×3 rotation matrix; and λ is a scale factor.

The location of the projection centre and the rotations as a function of time can be computed by means of the inertial satellite orbit and the satellite velocity. Instead of this strict computation, the relationship between the ground coordinates and time is often interpolated between the given orbit ephemeris and attitudes, usually by second-order polynomials such as $X_{0_t} = X_o + a_1 t + b_1 t^2$.

The direct use of the collinearity equation requires level 1A type or basic imagery. Level 2A type images, projected to a constant object height and into the national map projection require a modified mathematical model or a computation of the original image coordinates from the level 2A type images.

Several software solutions are available for the orientation of images with constant view direction across orbit, described for example in: Konecny *et al.* (1987), Gugan and Dowman (1988), Kratky (1989), Baltsavias and Stallmann (1992), Jacobsen (1994), Fritsch and Stallmann (2000), Toutin (2000), Hattori *et al.* (2000), Dowman and Michalis (2003) and Toutin (2004). Some of these programs are not limited to images with fixed view direction. The details of software which is used commercially are often not available.

Sometimes the full orientation information is not available, or for some sensors the information in the header file is not self-explanatory; for this reason, in the Hannover programme BLASPO the projection centre coordinates from header files are not used, but the general orbit information about inertial orbit ellipse and inclination is used instead. With the attitude information (for SPOT given as look angles ψ_X ψ_Y in the metadata) and at least one ground control point, the right ascension (Figure 6.13), corresponding to the scene centre, can be computed. With at least three control points, the six described orientation unknowns, which are required for precise results, can be determined. For any ground point, the right ascension will be different because of Earth rotation. With this solution, even a stereo strip of four scenes in both the used orbits has been adjusted with just four GCPs, reaching sub-pixel accuracy in all three coordinate components at independent check points (ICPs). For longer stereo scene combinations in the latitudes of Europe, the convergence of the orbits reduces the lateral scene overlap significantly (see also Figure 4.1a).

For geolocation of low resolution reconnaissance images taken with nadir view, 2D polynomial solutions should be avoided. The possible inclined images, for example from SPOT, IRS-1C/1D or KOMPSAT-1, cause horizontal shifts of points, depending on the ground height. With unequal spacing of control points and varying height, large positional errors can be caused by 2D polynomial models. The nearly parallel view direction leads to position errors of $dL = dh \cdot \tan i$, where i is the incidence angle, dh is the height difference with respect to a reference plane and dL is position error (Figure 6.15). Height errors are caused by interpolation between neighbouring control points; this means that the distance between control points must be small enough to estimate the height model. The determination of a large number of ground control points is expensive or at least time consuming.

Because height is ignored, a simple 2D polynomial model, also known as "rubber sheeting", can only be used for small, flat areas, but even here there is a potential danger – errors in control points are hidden and cannot be easily identified. Of course, a simple polynomial fitting can be used for any type of image, but is not recommended.

6.3.2 Images from agile satellites

6.3.2.1 Image geometry and geometric reconstruction

The orientation of images from agile satellites is more complex – the rotation of the satellite may be changed continuously during imaging, not only because of an

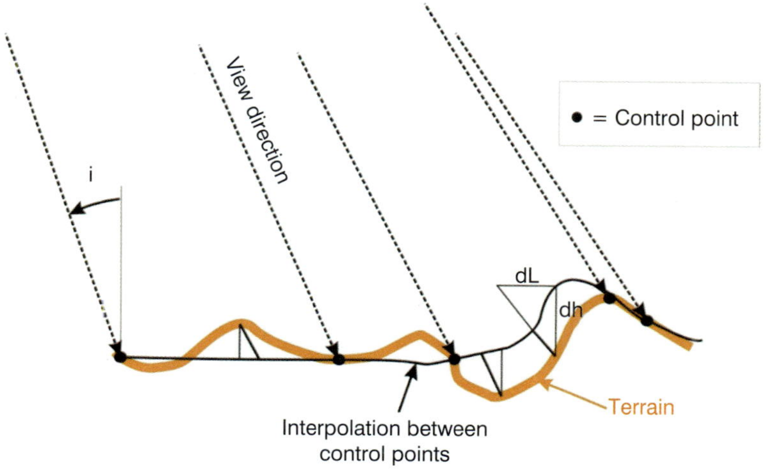

Figure 6.15 Positional errors caused by 2D polynomial fitting to control points.

asynchronous imaging mode, but also for fitting to a specified area on the ground (Figure 6.16). As shown earlier in Figure 5.5, most agile satellites can scan the area against the orbit direction. All three attitude angles may be changed, or even just the yaw by approximately 80°, to scan across the orbit (Figure 6.16b). In addition full information about the camera geometry, position and orientation is not always given for the high resolution commercial satellites. Instead, RPCs are provided to allow orientation to be determined using an alternative model (see Section 6.3.2.2).

Important progress in using optical space images came with IKONOS in 1999. Space Imaging tried to limit precise georeferencing and other value-added processing by the end user. They did this by not releasing details about inner and exterior orientation. Only the IKONOS Geo image products, formerly called CARTERRA Geo, are distributed; these are level 2A type images, projected to a plane of constant height and referenced to a national coordinate system. Such georeferenced images are not orthoimages which have been rectified to a digital elevation model. Orthoimages were 5.5 times more expensive than IKONOS Geo. Users tried to save money by rectifying IKONOS Geo images themselves. As mentioned in Section 6.3.1, 2D polynomial models only can be used in small, flat areas, requiring several control points. Also, other approximations (described in more detail later) have been used. Toutin and Cheng (2000) published the article "Demystification of IKONOS" in 2000, in which they discussed a 3D parametric model, implemented in PCI Geomatica, but without explaining details of the solution. The program CORIKON was developed in 2000 at the University of Hannover; this reconstructed the imaging geometry based on the available "nominal collection elevation" and "nominal collection azimuth", together with

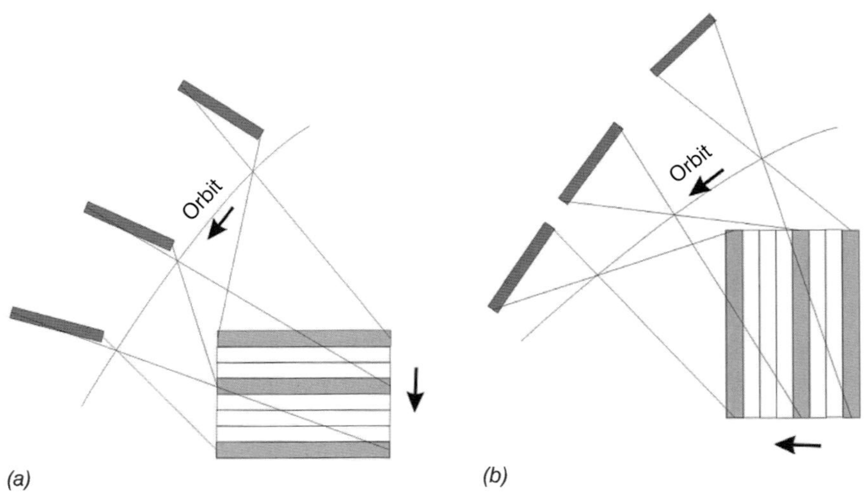

Figure 6.16 Possible scan directions of agile satellites: (a) rotation during orbit to national coordinate system; and (b) primary rotation about the z axis, plus fit to national coordinate system.

general orbit information about the IKONOS satellite [Jacobsen, 2002]. From the scene centre, or, in the case of OrbView-3, from the centre of the first line, the vector to the satellite is given by "nominal collection azimuth" and "nominal collection elevation" (Figure 6.17). The scene centre is given by its geographic coordinates. The orbit, given by orbit inclination and orbit ellipse, can be shifted with the right ascension to intersect with the vector from the scene centre to the satellite.

Based on the distance A in the scene, which may also have another orientation for other scan directions (see Figure 6.16), the distance B in the orbit can be calculated by following relationship

$$B = A \,.\, (\text{Earth radius} + \text{satellite height}) \,.\, (\text{slow - down factor} \,/\, \text{Earth radius}) \qquad (6.3)$$

Given the distance B in the orbit, the difference in time for imaging the scene centre can be computed, leading to a correction of the right ascension as a function of the Earth rotation. With this computation, the location of the projection centre for any individual scene line is known, leading to the direction from the image point in the level 2A type image to the object point passing through the projection centre. Internally in the program, transformations between inertial and Earth-centred Earth-fixed coordinates as well as between geographic and national coordinates have to be computed. With the height difference between the object point and the known height level of the reference plane for the level 2A type image, the ground coordinates X and Y of the

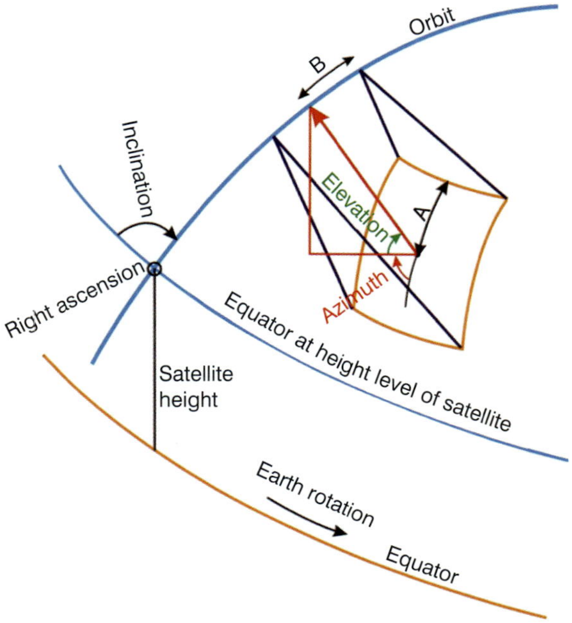

Figure 6.17 Geometric reconstruction of level 2A type images.

object point based on the direct sensor orientation can now be computed (Figures 6.18 and 6.19). This step is also called "terrain relief correction".

In the early years of IKONOS, no information about the height level of the reference plane was given in the metadata and it had to be estimated by the program user. The average height level of the GTOPO30 height model in the scene area was used, later confirmed to be a very good estimation. A 100 m error of the reference height (Figure 6.20) for the level 2A type image causes a scale error of 1.6 m in line direction

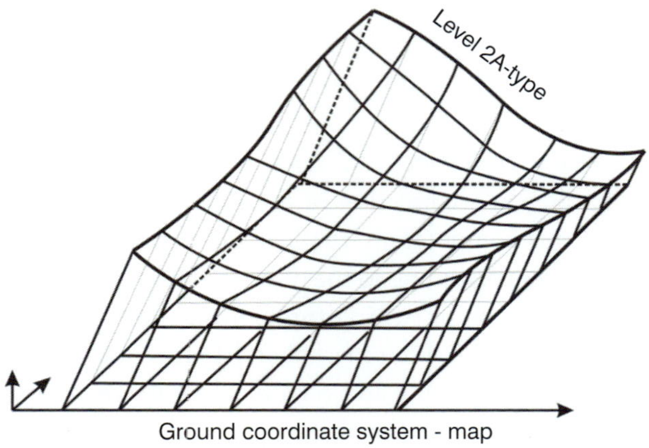

Figure 6.18 Geometry of level 2A type image (e.g. IKONOS Geo, OR standard).

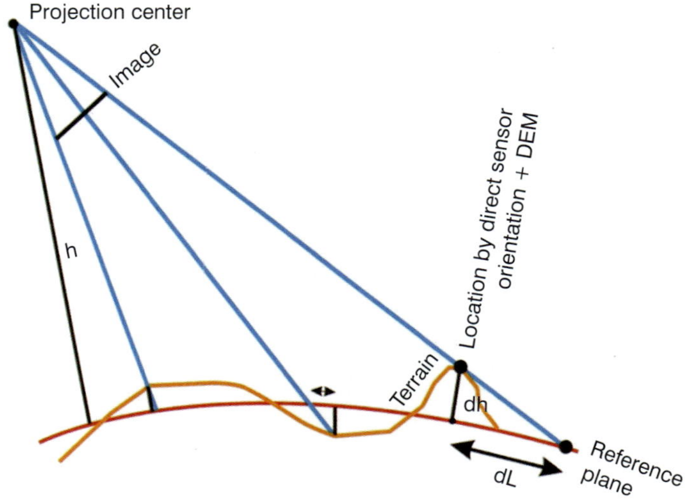

Figure 6.19 Georeferencing by direct sensor orientation based on the position of a point in a level 2A type image, the corresponding projection centre and the height difference between the ground point and the reference plane (terrain relief correction) – see Section 1.2.

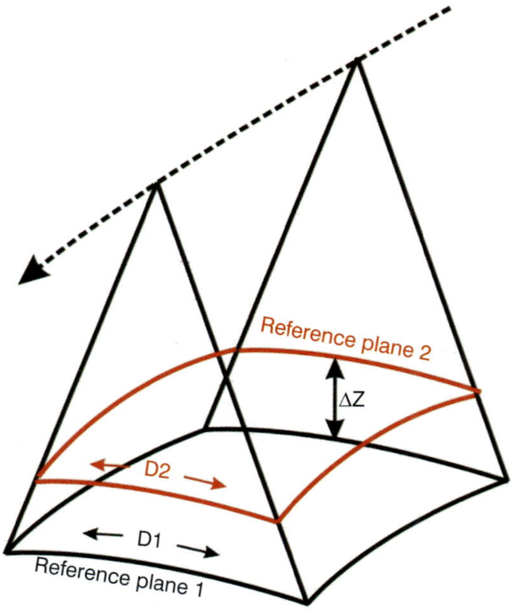

Figure 6.20 Influence of reference plane uncertainty on distance in a level 2A type image.

over a scene size of 11 km; for IKONOS this cannot be neglected. The scale error occurs only in the line direction; the sample direction is not influenced, leading to an affine deformation of the scene.

The geometric modelling of level 2A type images in the Hannover program CORIKON is done in two steps: first, the terrain height is used for a correction of the level 2A type image (Figure 6.19) (terrain relief correction); second, the horizontal position is improved by means of ground control points, named known as "bias correction". After the first step, which is carried out with just the control point height, the geolocation is available at the accuracy level of the direct sensor orientation. For IKONOS this is at the level of a standard deviation of 4 m in the WGS84 ITRF datum. In relation to a national coordinate system, the uncertainty of the national datum has to be added. The bias correction without knowledge of the height of the reference plane has, in theory, to be made with a 2D affine transformation including six unknowns, given by the equations

$$X = a_1 + a_2 . x + a_3 . y$$
$$Y = b_1 + b_2 . x + b_3 . y \tag{6.4}$$

where X and Y are ground coordinates, x and y are image coordinates and a_1–a_3 and b_1–b_3 are the coefficients of the affine transformation. If the height level of the reference plane for the generation of level 2A type images is given in the metadata, a simple shift with two unknowns may be sufficient for IKONOS, but not for some other satellite images.

In the Hannover program CORIKON the individual affine parameters are analysed for significance and mathematical probability to determine their values. With at least three control points, not located in a straight line, the six parameters of the affine transformation can be computed. Parameters not significant at a probability level of 68% should be removed from the adjustment: they are not required and do not improve the result. CORIKON also allows the determination and improvement of nominal collection of elevation and azimuth, but this requires a good 3D distribution of the ground control points and is only justified if no metadata are available.

Figure 6.21 shows the root mean square error (rmse) at independent check points of IKONOS orientations over the test field Zonguldak in Turkey, as a function of the number of GCPs used, computed with bias correction by a shift and a 2D affine transformation. It can be seen that the fit to independent check points could not be improved by bias correction with the six unknowns of a 2D affine transformation, compared to a bias correction with just two unknowns of a shift. With a limited number of control points, a simple shift after terrain relief correction led to slightly better results than the affine transformation. This can only be explained by small discrepancies at some of the control points, which have a smaller influence in the case of a simple shift. IKONOS has a very good inner accuracy; except for scenes without information about the height of the

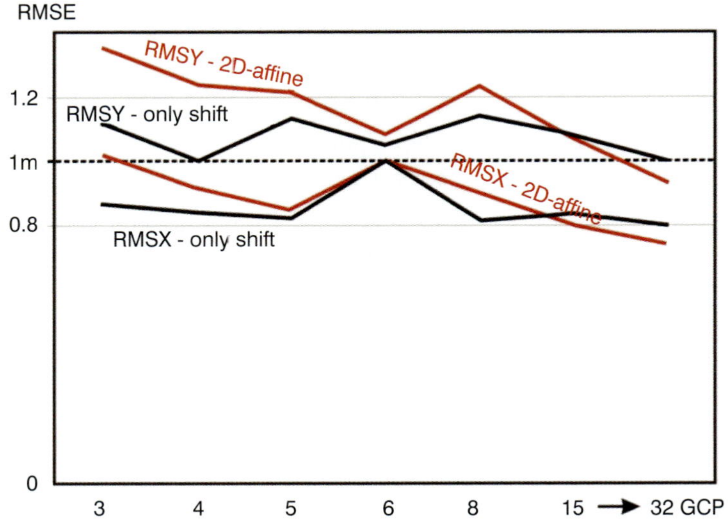

Figure 6.21 Root mean square errors (rmse) at independent check points as a function of the number of ground control points used, after orientation of an IKONOS scene by geometric reconstruction and bias correction with shift and 2D affine transformation.

reference plane, only shifts in X and Y directions have been required, so in theory with just one control point, sub-pixel accuracy can be achieved. In general, the results from several orientations do not justify a bias correction of IKONOS scenes by 2D affine transformation. Apart from WorldView-1, the affine transformation improved the georeferencing of the scene and most of the affine parameters have been shown to be significant.

The orientation of agile satellites can also be directly based on orbit positions and attitude data if it is provided, which may be improved by means of ground control points. The mathematical solution may be similar or even identical to the orientation of images with constant view direction across the orbit. The orientation of stereo combinations from stereo satellites also can be supported by common adjustment using tie points in addition to ground control points. The Space Application Centre (SAC) of the Indian Space Research Organisations (ISRO) developed a stereo strip triangulation (SST) technique, improving the attitude parameters, influenced by satellite motion, by means of ground control points and tie points between the forward- and the backward-looking camera images of Cartosat-1 [Gupta *et al.*, 2008]. A linear correction model for correcting the measured attitude values is used. With this method long stereo strips can be oriented. A Cartosat-1 stereo strip of 670 km length could be oriented with just eight ground control points. A root mean square error of 1.5 to 2.5 GSD was achieved. Over a strip length of 1533 km, with just two ground control points, root mean square errors of 4 GSD were obtained.

ETH Zurich (Swiss Federal Institute of Technology) included in their SAT-PP software system, besides others, a sensor trajectory model for the orientation and calibration of ALOS PRISM, which has advantages for the orientation and calibration of the ALOS PRISM forward-, nadir- and aft-looking scenes with just few ground control points, compared to an improvement of the direct sensor orientation by a piecewise polynomial model (PPM) [Kocaman and Gruen, 2007]. For the piecewise correction of the orbit and attitude data, more control points are required. Nevertheless, with both methods a sensor calibration and orientation is possible, achieving sub-pixel accuracy for a standard scene size with triple image coverage.

The Hannover program BLASPO, developed for the orientation of space images with constant view across the orbit, can be used also for the adjustment of level 1A type images from agile satellites [Jacobsen and Passini, 2003]. To reach the full accuracy potential of QuickBird basic imagery, a few additional parameters are required if the complete information about direct sensor orientation is not used.

6.3.2.2 Sensor-oriented rational polynomial coefficients (RPCs)

When IKONOS-2 data became available, Space Imaging decided not to distribute the orientation information. After the development of programs able to compute the satellite image orientation without detailed information, they made sensor-oriented RPCs available and extended the metadata.

The geometric relationship between image and ground coordinates can be replaced by a 3D interpolation in the object space with polynomials based on the sensor geometry and direct sensor orientation [Grodecki, 2001]:

$$x_{ij} = \frac{P_{i1}(X,Y,Z)_j}{P_{i2}(X,Y,Z)_j} \qquad y_{ij} = \frac{P_{i3}(X,Y,Z)_j}{P_{i4}(X,Y,Z)_j} \tag{6.5}$$

where x_{ij}, y_{ij} are the normalised image coordinates; X, Y, Z are the normalised ground coordinates; and P is a polynomial function – usually third order. The polynomial functions normally used are:

$$\begin{aligned} P_{i1}(X,Y,Z)_j &= a_1 + a_2 .Y + a_3 .X + a_4 .Z + a_5 .Y.X + a_6 .Y.Z + a_7 .XZ + a_8 .Y^2 \\ &+ a_9 .X^2 + a_{10} .Z^2 + a_{11} .XYZ + a_{12} .Y^3 + a_{13} .YX^2 + a_{14} .YZ^2 + a_{15} .Y^2 X \\ &+ a_{16} .X^3 + a_{17} .XZ^2 + a_{18} .Y^2Z + a_{19} .X^2Z + a_{20} .Z^3 \end{aligned} \tag{6.6}$$

The coefficients a_1–a_{20} are provided in the metadata, and the polynomials P_{i2}, P_{i3} and P_{i4} are of a similar form, making a total of 80 coefficients. Geographic coordinates are used as ground coordinates. To avoid numerical problems, all values are normalised with values between −1.0 and +1.0, in respecting offsets and scale factors for the object and image coordinates.

The RPC values are computed based on the direct sensor orientation and known interior orientation, which is determined by positioning with a global navigation satellite system like GPS and attitude information from star sensors and gyros, together with a system calibration. A cube of object points with a specified number of layers has to be generated around the scene area (Figure 6.22). With the information from direct sensor orientation and the system calibration, the image coordinates are computed for any ground coordinate in this cube of points. The RPCs are adjusted with a large number of corresponding 3D object and image coordinates, well distributed throughout the scene.

Usually the image coordinates are expressed by third-order polynomials of the geographic coordinates X, Y, Z (Equation 6.6). The sensor-oriented RPC solution is an approximation, but with the high number of coefficients the loss of accuracy is negligible compared to a rigorous model for usual scene size [Dowman and Dolloff, 2000]. Similar to the geometric reconstruction, the sensor-oriented RPCs are used for the terrain relief correction (Figure 6.19). This should be followed by bias correction; this is a 2D transformation to ground control points, as in the case of geometric reconstruction. As in the case of the general results, the bias correction of the sensor-oriented RPC solution is nearly identical to that of the geometric reconstruction. Hanley and

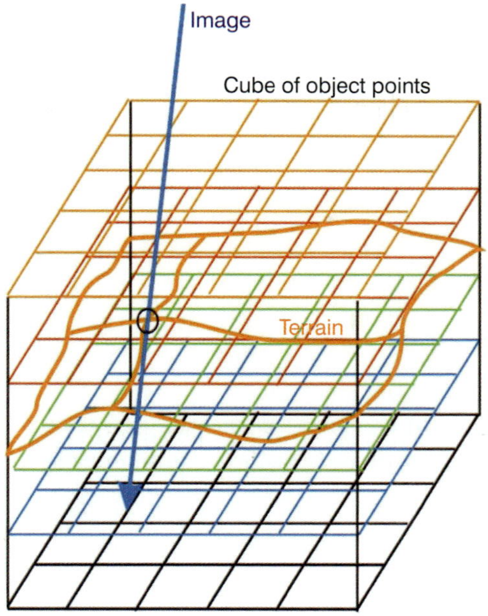

Figure 6.22 Determination of RPCs with a cube of object coordinates and corresponding image coordinates.

Fraser (2001) and Fraser *et al.* (2002) demonstrated the effectiveness of bias correction with a 2D affine transformation for IKONOS data. The comparison of the bias correction by a shift or a 2D affine transformation in the Zonguldak test area shown in Figure 6.21 has also been made for the RPC solution, leading to almost the same result. For other sensors, it was similar for both of the rigorous orientation methods.

The requirement of the 2D affine transformation for bias correction of a QuickBird scene in the Zonguldak test area can be seen in Table 6.2, which shows the root mean square discrepancies of 40 control points. By bias correction with a 2D affine transformation, the root mean square discrepancies of the control point X coordinates are reduced by a factor of around three, compared to a simple shift. The improvement in the Y direction (north direction) is smaller, but also significant. In this case the orientation by sensor-oriented RPC gave better results than the orientation by geometric reconstruction. When an additional unknown for correction of the asynchronous imaging mode slow-down factor is added, the results of both orientation methods became closer to each other.

The RPC solution is simple to handle for level 2A type images. Such scenes are georeferenced, including approximate values of the ground coordinates, which can be used as input for Equation (6.5). Level 1A images or basic images, which are close to the original images, are not directly accommodated by the RPC solution; in this case an iterative method is required instead. In the Hannover program RAPORIO, with the

Table 6.2 Root mean square discrepancies at 40 control points, showing the effect of bias correction, for QuickBird over the Zonguldak test area

	Sensor-oriented RPC		Geometric reconstruction	
	$rmse_X$ (m)	$rmse_Y$ (m)	$rmse_X$ (m)	$rmse_Y$ (m)
Shift	1.63	0.57	1.88	0.88
2D affine	0.38	0.51	0.68	0.63
2D affine + slow-down correction	–	–	0.44	0.59

given ground coordinates of the scene corners, a perspective transformation of the image positions to ground positions is made according to the equations:

$$X = \frac{a_1 + a_2 . x + a_3 . y}{c_1 . x + c_2 . y + 1}$$
$$Y = \frac{b_1 + b_2 . x + b_3 . y}{c_1 . x + c_2 . y + 1} \tag{6.7}$$

where X and Y are ground coordinates, x and y are image coordinates and a_1–a_3, b_1–b_3 and c_1–c_3 are coefficients of the perspective transformation. The ground coordinates are then used as input to Equation (6.5) but will not lead to the exact original image positions. After two to three iterations, however, the values converge to the optimal solution.

6.4 Approximate orientation methods

6.4.1 Terrain-dependent RPC solution

The relationship of scene to object coordinates can also be approximated by RPCs just based on control points, without any information about the imaging geometry and any information from metadata. This method is known as the "terrain-dependent" RPC solution. It is not possible to determine 80 orientation unknowns of the third-order polynomials: this would require at least 40 3D well-distributed ground control points per scene, so only a selected number of RPCs are computed. The number of chosen unknowns is dependent on the number of control points and their 3D distribution.

Geometric problems of this method cannot be controlled simply by the residuals at the ground control points. Some commercial programs offering this method do not use any statistical check for high correlation of the unknowns, making the correct handling of the terrain-dependent RPC solution very dangerous. A test with IKONOS data and a commercial program using this method showed residuals at control points below 1 m and the software did not indicate any problems. However, in an area within the range of the control points, at check point discrepancies of up to around 50 m, and outside the range of the control points, discrepancies exceeding as much as 500 m appeared (Figure 6.23). In this case the control point distribution is not optimal, even though with exactly the same image and ground control points, the orientation by geometric reconstruction and the sensor-oriented RPC solution resulted in root mean square errors (rmse) at independent check points in the region of ±1.0 m. The terrain-dependent RPC solution cannot be controlled simply by analysing discrepancies at ground control points: at least the correlation between the parameters has to be checked, but this is not included in commercial versions that use this method. This method is therefore not a serious solution and should never be used. In general, available orientation information should be used to reduce the required number and distribution of ground control points.

6.4.2 3D affine transformation

A simple method for approximating the relationship between image and ground coordinates for space images having a narrow field of view, the 3D affine transformation was developed by Okamoto (1981) and investigated by the University of Melbourne for very high resolution optical space imagery [Fraser *et al.*, 2002; Hanley *et al.*, 2002]. The 3D affine transformation (Equation 6.8 below) expresses the relationship between ground coordinates and image coordinates without any pre-information about the sensor orientation. It is the mathematical model for a parallel projection, which is valid only for image planes that are approximately parallel to the ground coordinate system and for a narrow field of view, because it neglects the perspective geometry in CCD-line direction (Figure 6.24). For the typically inclined view direction of very high

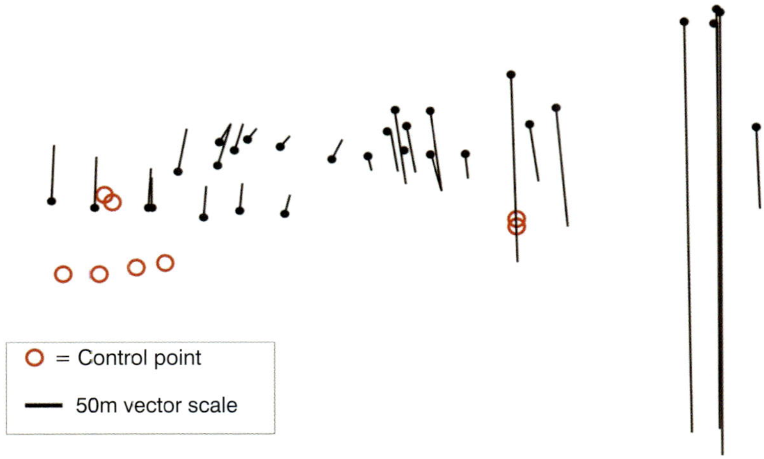

Figure 6.23 Discrepancies at independent check points from an IKONOS scene orientation by a terrain-dependent RPC solution, highlighting the dangers of using a poorly controlled solution.

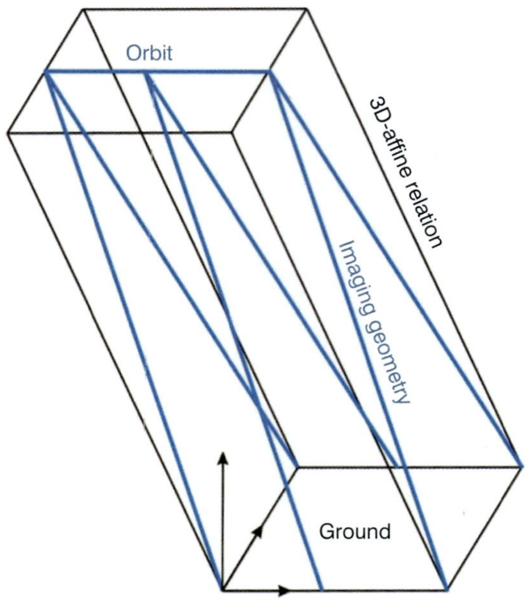

Figure 6.24 Geometric model of a 3D affine transformation in relation to the perspective image geometry in the CCD-line direction.

resolution imagery, the technique can only be used for level 2A type images (Geo or OR standard) and for sensors that utilise only a limited slow-down factor for asynchronous imaging.

The 3D affine transformation allows a rotation and independent change of scale in *x*, *y* and *z*. It additionally makes a correction to the height, depending on the elevation of the point transformed, as shown in Equation (6.8). The influence of the Earth's curvature in the orbit direction is approximately compensated for by the curvature of the orbit. For an IKONOS scene with 20° incidence angle, the effect of Earth curvature in the line direction is limited to 0.28 m in the root mean square error, which is acceptable for 1 m GSD. But, in the case of larger object height variation, positional errors cannot be avoided. For the 3D solution, 3D distributed control points are required to allow the view direction to be determined.

The formula for a 3D affine transformation is given by

$$\begin{aligned} x_{ij} &= a_1 + a_2 . X + a_3 . Y + a_4 . Z \\ y_{ij} &= b_1 + b_2 . X + b_3 Y + b_4 . Z \end{aligned} \tag{6.8}$$

where *X*, *Y* and *Z* are ground coordinates, *x* and *y* are image coordinates, and a_1–a_4 and b_1–b_4 are the coefficients of the transformation. The perspective geometry in the line direction and a slow-down factor of asynchronous imaging, as well as scanning across the orbit, can be taken into account with up to 12 terms using Equation (6.9) – called the "extended 3D affine transformation" [Jacobsen, 2008].

$$
\begin{aligned}
x_{ij} &= a_1 + a_2 \,.\, X + a_3 \,.\, Y + a_4 \,.\, Z + a_5 \,.\, XZ + a_6 \,.\, YZ \\
y_{ij} &= b_1 + b_2 \,.\, X + b_3 \,.\, Y + b_4 \,.\, Z + b_5 \,.\, XZ + b_6 \,.\, YZ
\end{aligned}
\tag{6.9}
$$

Footprints of level 1A-type images (original or basic images) after terrain relief correction must not have parallel sides caused by the projected pixel size changing with the incidence angle (Equations 4.1 and 4.2), especially in the case of an asynchronous imaging mode (Figure 6.25). (However, an image could have parallel sides if the camera axis was vertical.) Parallel sides are required for the 3D affine and extended 3D affine transformation (Equation 6.9), so a further extension by two terms to the following equation with 14 unknowns has to be made for level 1A-type or basic images, requiring at least seven 3D well-distributed control points [Jacobsen, 2008]

$$
\begin{aligned}
x_{ij} &= a_1 + a_2 \,.\, X + a_3 \,.\, Y + a_4 \,.\, Z + a_5 \,.\, XZ + a_6 \,.\, YZ + a_7 XX \\
y_{ij} &= b_1 + b_2 \,.\, X + b_3 \,.\, Y + b_4 \,.\, Z + b_5 \,.\, XZ + b_6 \,.\, YZ + b_7 XY
\end{aligned}
\tag{6.10}
$$

The 3D affine transformation and its extension require the determination of the view direction by means of control points located at different height levels. This is not so critical as in the case of the terrain-dependent RPC solution, in which the projection direction may change locally. For the standard 3D affine transformation (Equation 6.8),

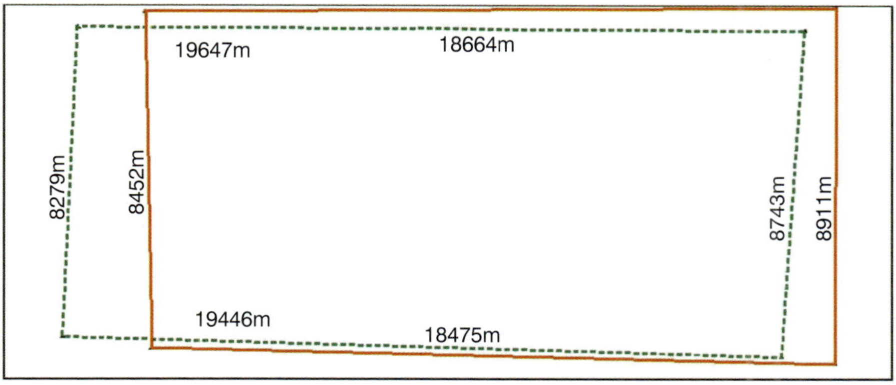

Figure 6.25 Footprints of an OrbView-3 basic imagery stereo pair; opposite sides not parallel.

only one view direction has to be determined, while the extended 3D affine transformation also requires information about the changing view direction. Even in mountainous areas, the ground control points are mainly located within a small height range because control points on the tops of mountains cannot be reached as easily as points in the valleys. In such a case, points located on the tops of mountains are in an area of extrapolation, out of the 3D range of control points. Extrapolated points may have a limited accuracy. In the mountainous area of Zonguldak the geometric potential of several space images has been investigated based on precise ground control points. The orientation of an IKONOS scene with a 3D affine transformation was made using four control points having heights between 3 and 186 m. Check points reach elevations of up to 420 m. The 3D affine transformation has eight unknowns, so no over-determination exists for four control points, but the control points had been checked in advance with a higher number of control points, so satisfactory results were expected under this condition.

At independent check points, root mean square errors in X of 1.91 m and in Y of 18.53 m were reached (Figure 6.26). This is not an acceptable result for IKONOS scenes with 1 m GSD. Using the same control points, root mean square errors at check points in the region of 1 m have been achieved with a geometric reconstruction and bias-corrected sensor-oriented RPC solution. Figure 6.26 gives an explanation for the poor results: three control points are located almost on a straight line at heights below

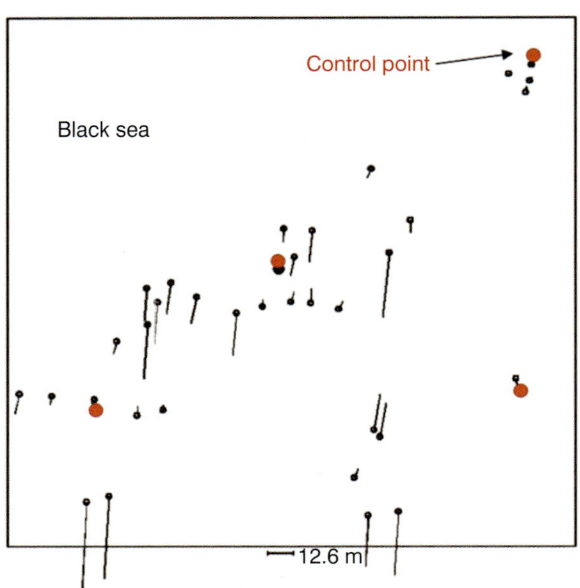

Figure 6.26 Analysis of IKONOS orientation, using a 3D affine transformation in a mountainous area.

10 m, and the fourth, in the south-east corner, has a height of 186 m, so the four control points describe a plane that is nearly tilted, and all points not located in the vertical direction close to this tilted plane show larger discrepancies at check points. The unknowns of the 3D affine transformation are correlated with each other, between 0.95 and 0.98, demonstrating the requirement for using control points in a cube around the area to be handled with the 3D affine transformation.

IKONOS has a field of view in the line direction of 0.92°, so the error in the mathematical 3D affine model of approximating the bundle of rays from the orbit to the ground by parallel lines is limited to a maximum of 0.46° of the ray direction, leading to a maximum error of 0.8 m in the horizontal position for points with height differences of 100 m with respect to the average plane of control points. In a mountainous area with larger height differences, this is not acceptable for an image with 1 m GSD, limiting the use of the 3D affine transformation for IKONOS to areas with gentle height variations. QuickBird has approximately twice as many pixels in the line as IKONOS, as well as a lower flying height, resulting in 2.13° field of view in the line direction. The corresponding maximum error of the ray direction in the line of 1.07° is 1.9 m over 100 m difference in height; this is not acceptable in a scene with 0.62 m GSD. In addition, QuickBird operates with a slow-down factor of 1.6 for the asynchronous imaging mode, leading to a maximum error of the ray inclination of 0.6° in the sample direction for a scene length of 16.8 km, corresponding to a location error of 1.1 m for 100 m difference in height. For this reason, the standard 3D affine transformation is limited to very flat terrain for scenes with a comparable field of view to QuickBird. The extended 3D affine transformation (Equation 6.9), using 12 unknowns, respects a continuous change of the ray direction over the scene, including the perspective geometry in the line direction, but 12 unknowns require at least six control points and they must be well distributed in all three coordinate components.

6.4.3 Direct linear transformation

Another approximate orientation solution, not using any pre-information about the imaging geometry, is the direct linear transformation (DLT) given by:

$$x_{ij} = \frac{L_1 . X + L_2 . Y + L_3 . Z + L_4}{L_9 . X + L_{10} . Y + L_{11} . Z + 1}$$
$$y_{ij} = \frac{L_5 . X + L_6 . Y + L_7 . Z + L_8}{L_9 . X + L_{10} . Y + L_{11} . Z + 1} \tag{6.11}$$

With 11 parameters, it expresses the inner and exterior orientation of a perspective image and is designed for close-range applications, where it works well with ground control points having a large range of coordinates in the direction of the optical axis.

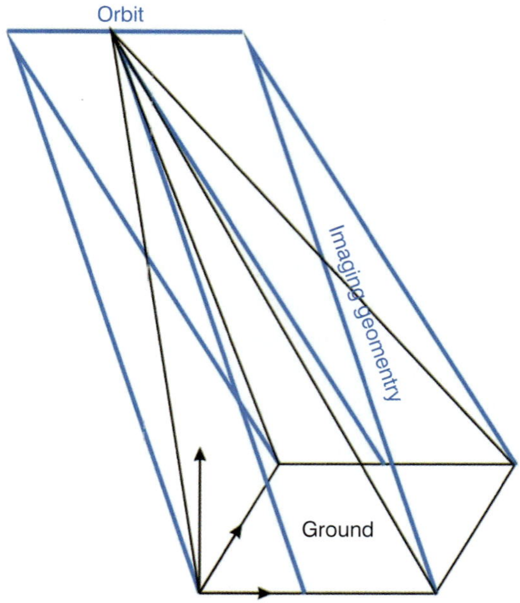

Figure 6.27 Geometric model of a direct linear transformation (DLT), showing the imaging geometry (blue) and the DLT (black).

The DLT does not respect the close-to-parallel projection in the orbit direction, as shown in Figure 6.27. For the 11 unknowns at least six 3D well-distributed control points are required.

Even when the DLT is used for aerial application, with quite a large field of view, high correlations of the DLT unknowns are very often created. This is more critical for the small field of view of space images. Once again, commercial programs offering the DLT for space application do not analyse correlations and give no warnings to the users. The IKONOS data set shown in Figure 6.28 includes six control points, well distributed in elevation between 6 and 420 m, but the root mean square discrepancies at independent check points reached 2.4 m, compared to the sub 1 m accuracy indicated by orientation with geometric reconstruction or a bias-corrected sensor-oriented RPC solution. A DLT with all 40 control points generated a correlation matrix in which, out of the 11^2 values of the covariance matrix, 18 values other than the diagonal elements were listed with correlation coefficients of 1.000 (that means they exceeded 0.9995). With such correlations, the result has a random nature. The DLT is very sensitive and fails in flat terrain; furthermore, in mountainous terrain the errors caused by the approximate mathematical model are not acceptable, so in general the DLT cannot be recommended.

The standard 3D affine transformation has errors in the mathematical model similar to the DLT: parallel projection instead of close-to-parallel projection in one direction. With 8 unknowns the 3D affine transformation usually leads to stable results; this

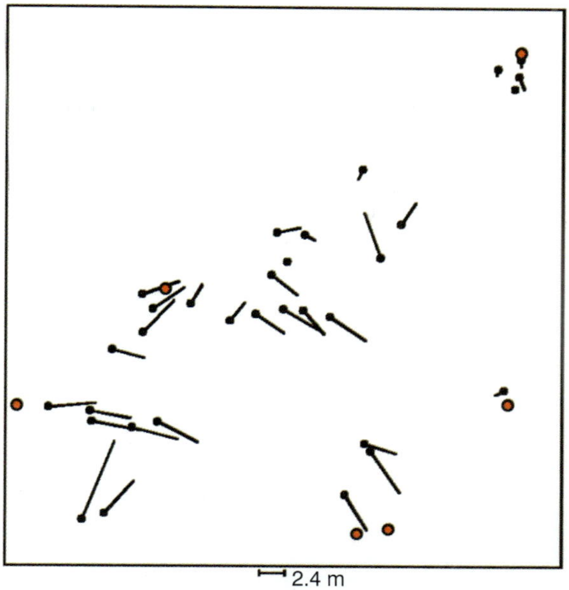

Figure 6.28 Discrepancies at independent check points after IKONOS scene orientation by DLT based on six control points (red circles).

is often not the case for the 11 unknowns of the DLT. For this reason, the 3D affine transformation can be used for a rough check of the data set before strict solutions are used for the final results.

6.5 Comparison of orientation methods

6.5.1 Issues in assessing accuracy

An analysis of the accuracy, which can be carried out with the different orientation methods and the different images, shows that the ground control point quality is a dominant component. The accuracy of the ground control points today is not the limiting factor: in most cases the survey will be done by relative GPS positioning, accurate enough for optical satellite images. The major problem is caused by the point identification and specification in the images. Optimal points are symmetric and have sufficient contrast and diameter. Traffic circles are very good, but rare.

Figure 6.29 shows the grey level profile through an edge and a symmetric point. If the grey value shoulder for the identification of an edge or corner position is changed, the position of the edge or corner is shifted, while in the case of a symmetric point the grey value shoulder for the point identification has no influence on the position.

Figure 6.30(a) shows an optimal symmetric element, usable as a control point. Figure 6.30(b) shows a point which is not well defined – the corner of this point was

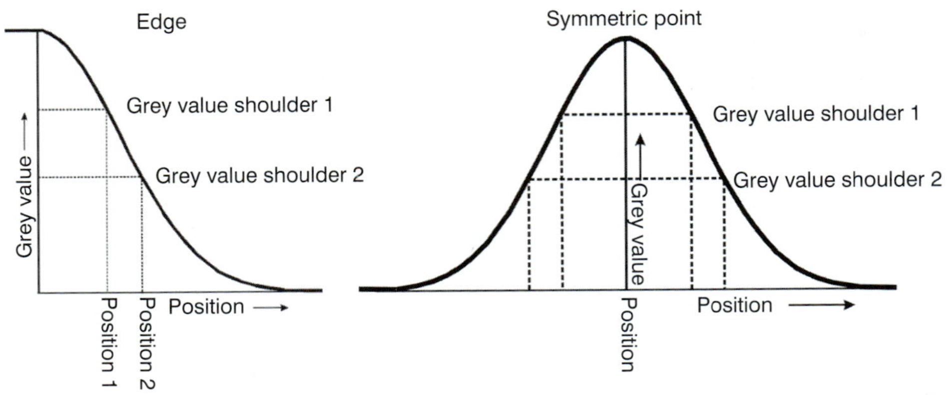

Figure 6.29 Grey value profile through an edge and a symmetric point.

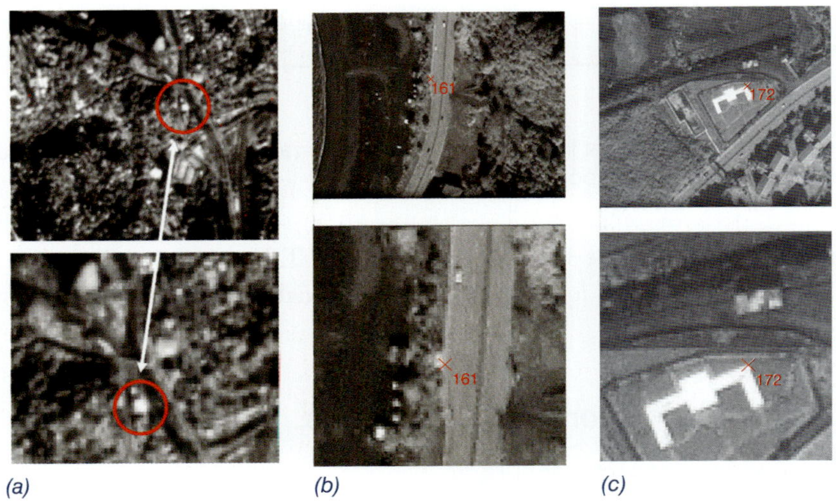

Figure 6.30 Comparison of images of control points: (a) optimal symmetric point; (b) poorly defined point; and (c) an extreme corner position.

used and the magnification indicates problems with the exact specification of the location. In Figure 6.30(c), the corner position of a building with strong contrast was used. Such a corner point changes the location in the image if the brightness or the contrast is changed, as is obvious in Figure 6.31(a). The diameter of the point in Figure 6.31(a) clearly varies by just a change in the image brightness, so a corner position at such a point shifts, depending on the brightness and contrast. Such a shift may exceed the size of a pixel – too much for a positioning with sub-pixel accuracy.

Part of a QuickBird image is shown in Figure 6.31. In this case the geometric potential of a QuickBird scene was checked using control points digitised from US Digital

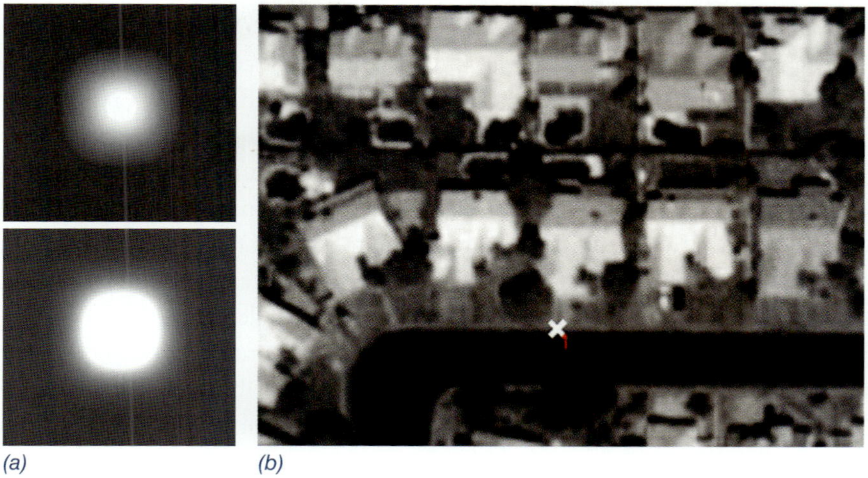

Figure 6.31 Examples of unsuitable control points: (a) the same point shown with different brightness; and (b) a point which is not optimal, selected on a grey value edge.

Orthophoto Quarter Quads (DOQQs), having 1 m pixel size [Jacobsen and Passini, 2003]. Initially, with 48 symmetric points, a root mean square error of 1.04 m of the QuickBird orientation was reached. This is very close to the estimated accuracy of the DOQQs, so the error component of the QuickBird scene must be smaller. Later 159 points, defined by intersections of garage exits and street edges, were used (Figure 6.31b), leading to a root mean square orientation value of 1.34 m. This corresponds to a root mean square difference of 0.84 m between both measurements, which can only be explained by the negative influence of the control points selected on edges. With such edge points, no sub-pixel accuracy can be reached. This is also obvious from Figure 6.32, in which the contrast enhancement causes approximately 1 pixel change in the corner position.

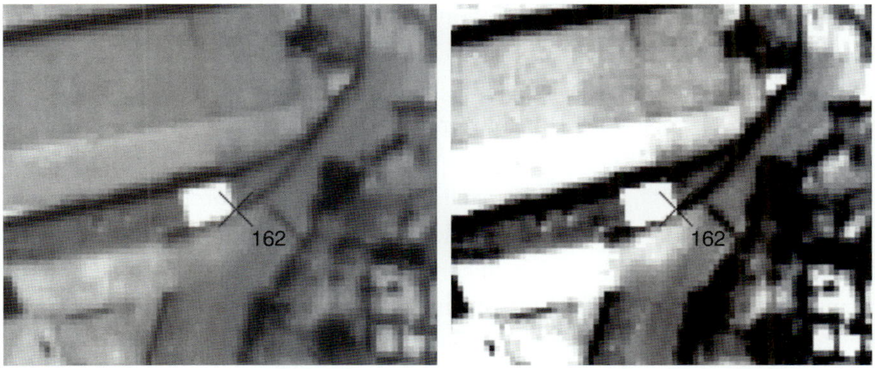

Figure 6.32 Influence of contrast enhancement on corner position, giving a shift of approximately 1 pixel.

6.5.2 Results from tests

In the Zonguldak test area the orientation of an IKONOS scene has been investigated with different orientation methods. Sensor-oriented RPCs and geometric reconstruction were used. The direct sensor orientation of IKONOS, based on GPS positioning, star cameras and gyros, is used and allows an orientation without control points. The systematic discrepancies of 0.7 m in X and 26.4 m in Y are dominated by the uncertainty of the local Turkish datum and not by the absolute IKONOS positioning quality. As mentioned above, for both rigorous orientation methods, after terrain relief correction, the bias correction by means of control points could be made by a simple shift; by 2D affine transformation to the control points, no accuracy improvement could be gained.

For an orientation by geometric reconstruction as well by bias-corrected sensor-oriented RPC solution it does not matter how many control points are used: the root mean square difference at independent check points is nearly constant and only slightly influenced by some random errors of the control and check points used. A slightly better accuracy was obtained with the RPC solution, but both methods show root mean square errors in pixels of up to sub-pixel order. The control points in the Zonguldak test area used for the IKONOS scene have height variations ranging from 6 to 420 m above mean sea level; this is not optimal for the approximate 3D affine transformation and DLT. The 3D affine transformation has 8 unknowns, requiring at least four control points; the extended 3D affine transformation, respecting the geometric property of space images, has 12 unknowns and the DLT 11 unknowns, both requiring at least six control points. These methods are very sensitive to the 3D distribution of the control points (as shown in Figure 6.24), but even with an optimal distribution, two control points more than the minimal number have to be used. Even with eight control points, the accuracy is below that of the rigorous methods (Figure 6.33).

The orientation of a QuickBird OR standard image in the test area of Zonguldak shows some similarities to the orientation of the IKONOS scene. For QuickBird, the bias correction after terrain relief correction has to be done by a 2D affine transformation. With just a shift for bias correction, using four control points the root mean square error at independent check points is 1.42 m instead of 0.62 m; the bias-corrected sensor-oriented RPC solution is similar to the geometric reconstruction. For this reason, Figure 6.34 starts with three control points that are required for determining the six unknowns of the bias correction. With the exception of the orientation using only three control points, the accuracy achieved with both rigorous orientation methods is nearly identical. The larger field of view of QuickBird limits the 3D affine transformation as well as DLT, even with all control points, to approximately 0.8 m instead of 0.5 m. The extended 3D affine transformation, respecting the perspective geometry in the line direction and the asynchronous imaging mode of QuickBird, results in approximately the same object point accuracy as the rigorous methods if at least ten control points are used. The reached accuracy of 0.5 m corresponds to 0.8 GSD. With especially the orientation methods using approximated mathematical models, two

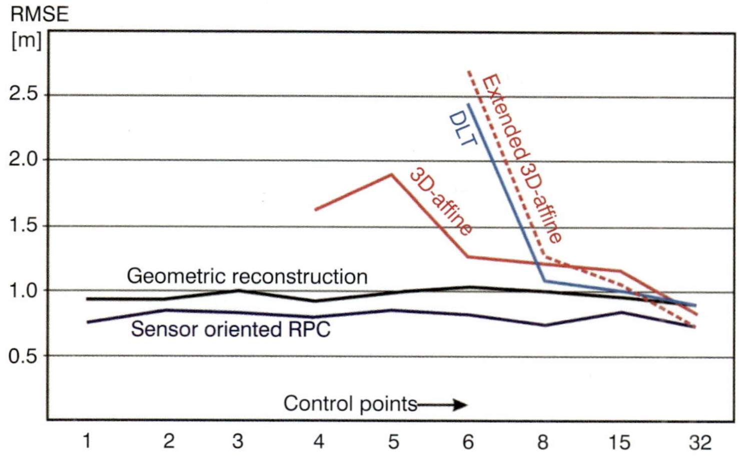

Figure 6.33 Results from the orientation of an IKONOS scene (1 m GSD) at Zonguldak (mountainous area). Root mean square discrepancies are given as an average of the X and the Y coordinate components at independent check points, as a function of the number of GCPs.

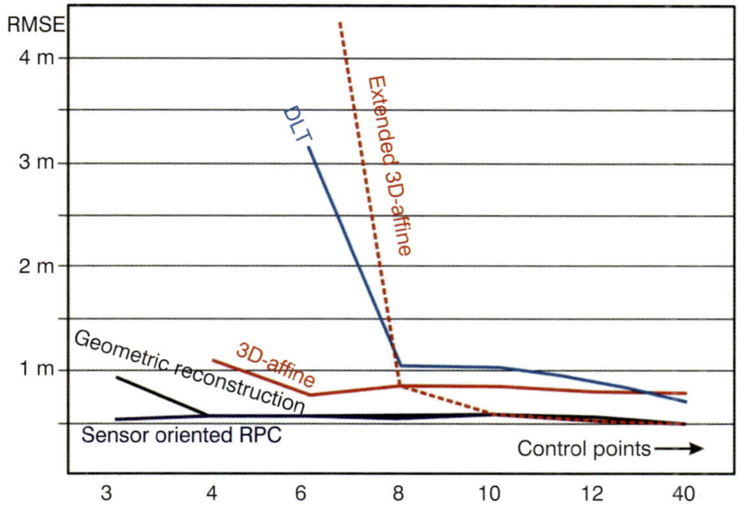

Figure 6.34 Orientation of QuickBird OR standard image (0.62 m GSD) of Zonguldak (mountainous); root mean square discrepancies at independent check points are shown as a function of the number of GCPs.

control points more than the minimum are required to achieve satisfactory results, independently of the general rule of using an over-determination for reliable results.

The orientation of OrbView-3 basic images using the same ground control points as for IKONOS and QuickBird in the Zonguldak test area, with the approximate orientation methods, highlights the greater difficulties of handling level 1A type imagery (Figure 6.35). The scene boundaries are not parallel, because of the changing incidence angle, enlarged by an asynchronous imaging mode, so the conditions for use of 3D affine transformation are not given. Even with 29 control points, the standard 3D affine transformation (Equation 6.8) is limited to 10 m accuracy in the average of both analysed images. Also, the extended 3D affine transformation (Equation 6.9) just leads to an improved value of 7 m. Only the 3D affine transformation extended to original images (Equation 6.10), having 14 unknowns, yields root mean square discrepancies of 2 m for 29 control points, which approaches the result of the RPC solution. The DLT is limited to 6 m accuracy. For 1 m GSD of OrbView-3, such results are not acceptable.

Only the RPC solution reaches root mean square differences of 1.3 m based on 29 control points; with 4 to 12 control points it is in the region of 1.6 m. This is still a significantly higher value than for other comparable optical space images. One of the reasons is the quality of the OrbView-3 images. OrbView-3 uses staggered CCD lines; neighbouring pixels are over-sampled by 50%, so images with 1 m GSD are generated from the projected pixel size of 2 m. This does not lead to the same image quality as images having 1 m projected pixel size. The control point measurement was more difficult than for IKONOS also having 1 m GSD. The pointing accuracy and the accuracy of the control point ground coordinates is indicated by the relative accuracy: the accuracy of one check point in relation to the neighbouring ones. For distances of up to 1 km, for which errors of the mathematical model do not play an important role, with IKONOS the relative accuracy is 0.75 m, while for OrbView-3 it is 1.0 m.

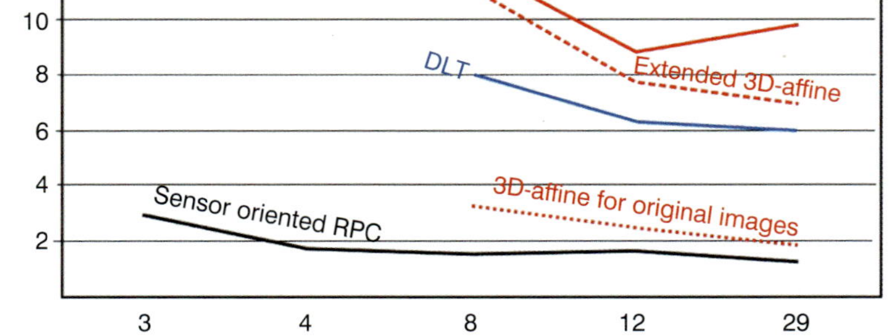

Figure 6.35 Orientation of OrbView-3 basic image (1 m GSD) of Zonguldak (mountainous area). Root mean square discrepancies at independent check points are shown as a function of the number of GCPs.

For QuickBird, with the same control and check points, it is 0.44 m corresponding to 0.71 GSD, showing the negligible influence of the control point ground coordinate accuracy. This indicates the more difficult pointing in OrbView-3 images, caused by the limited image quality.

The results of image orientation with a WorldView-1 OR standard scene in the city area of Istanbul, taken with a nadir angle of 33°, are shown in Figure 6.36. Control points are located only in an image tile of 8 km square, limiting the field of view used. The bias correction of the sensor-oriented RPC solution was possible, without loss of accuracy, by just a shift, allowing the orientation with only one control point. The root mean square discrepancies at independent check points are almost independent of the number of control points. Because of the small field of view used, with the extended 3D affine transformation and DLT a similar geometric quality to the sensor-oriented RPC solution was possible with at least 12 control points. The 3D affine transformation shows a strange behaviour as a function of the number of control points used; this can only be explained as a random result. The accuracy of the control point elevations was limited; a 33° nadir angle caused an error component in the view direction. The component perpendicular to the view direction is not influenced by the point elevation; for the sensor-oriented RPC solution it is between 0.48 m and 0.52 m, which corresponds to 1.0 GSD. For 33° nadir angle, the projected pixel size on the ground is 0.53 m × 0.64 m, so sub-pixel accuracy has also been reached with WorldView-1.

A GeoEye-1 stereo pair in Riyadh was tested using control points from aerial images, with a base-to-height ratio of 0.6. The following root mean square discrepancies were obtained: $rmse_X$ = 0.33 m, $rmse_Y$ = 0.40 m and $rmse_Z$ = 1.16 m. If the error

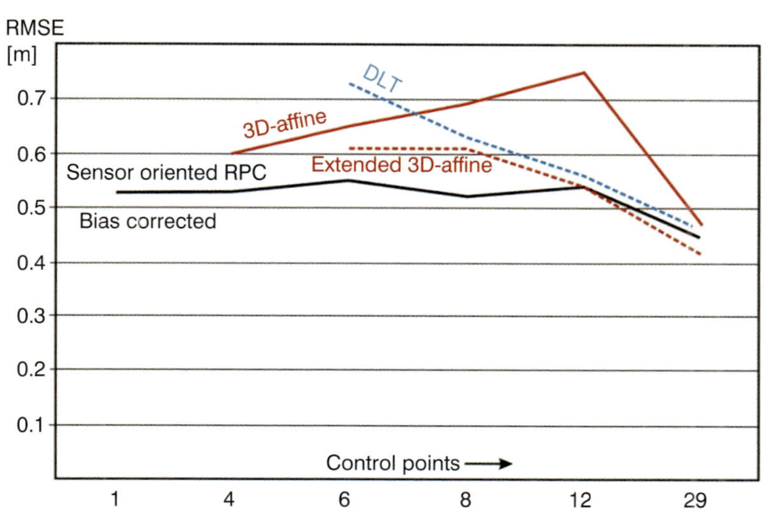

Figure 6.36 Orientation of WorldView-1 OR standard image (0.5 m GSD) of Istanbul. Root mean square discrepancies at independent check points are shown as a function of the number of GCPs.

component of the aerial images is respected, the vertical error component for GeoEye-1 is just 0.66. So sub-pixel accuracy is also possible with GeoEye-1 images. The orientation with a 3D affine transformation and DLT indicates a similar behaviour to the IKONOS images shown in Figure 6.33.

Concerning the orientation of IKONOS, QuickBird, OrbView-1, WorldView-1 and GeoEye-1 images, it can be concluded that the orientation quality of the bias-corrected sensor-oriented RPC solution, as well as the geometric reconstruction, is almost independent of the number of control points, but at least one control point above the minimum number should be used to obtain reliable results. The requirement of the bias correction after terrain relief correction by shift or 2D affine transformation depends on the type of satellite image: for IKONOS and WorldView-1, a shift is sufficient; for the other space images, a 2D affine transformation has to be used, requiring at least three control points. The extended 3D affine transformation (Equation 6.9) can be used for level 2A type images, provided enough control points are available. In general there is no justification in using approximate orientation methods that, with a limited number of control points, do not lead to the same geometric quality as the bias-corrected sensor-oriented RPC solution or the geometric reconstruction.

The orientation of single scenes of Cartosat-1 level 1A type stereo pairs by a bias-corrected sensor-oriented RPC solution leads to root mean square errors of 0.6 GSD to 0.8 GSD at independent check points, with precise and well-determined control points (Table 6.3). The 3D solution even improves the results at check points – to 0.5 GSD to 0.7 GSD in the horizontal component, demonstrating an error component of the control and check point height values [Jacobsen, 2007]. The standard deviation of the height can be calculated using Equations (1.8) and (1.9). The stereo satellite Cartosat-1 has a fixed base-to-height ratio of 0.63, so the standard deviation of the x-parallax is between 0.44 GSD and 0.54 GSD, demonstrating the high potential of this stereo system. With approximate orientation methods, no satisfactory results have been achieved with the level 1A type Cartosat-1 imagery.

In general sub-pixel accuracy of the object points can be reached with geometric reconstruction or a sensor-oriented RPC solution if well-determined control and check points are used (Table 6.4). If an accuracy of 1 GSD has not been achieved for the

Table 6.3 Root mean square differences at independent check points of Cartosat-1 (2.5 m GSD) stereo model orientation by bias-corrected sensor-oriented RPC solution

	3D solution			Orientation of single scenes (RMS average)	
Area	**$rmse_X$ (m)**	**$rmse_Y$ (m)**	**$rmse_Z$ (m)**	**$rmse_X$ (m)**	**$rmse_Z$ (m)**
Warsaw	1.33	1.94	1.57	1.94	1.57
Jordan	1.74	1.95	1.50	1.95	1.50

Table 6.4 Standard deviation of sensor orientation of high-resolution optical sensors at check points based on geometric reconstruction or a bias-corrected sensor-oriented RPC solution. σx/σy = root mean square average of X and Y ground coordinates.

	Level	GSD (m)	σ_X/σ_Y (m)	σ_X/σ_Y (GSD)
ASTER, Zonguldak	1A	15.0	10.8	0.7
KOMPSAT-1, Zonguldak	1A	6.6	8.5	1.3
SPOT, Hannover	1A	10.0	4.6	0.5
SPOT-5, Zonguldak	1A	5.0	5.2	1.0
SPOT-5, Zonguldak	2A	5.0	5.1	1.0
SPOT HRS, Bavaria	1A	5.0 × 10.0	6.1	0.7/1.1
IRS-1C, Hannover	1A	5.7	5.1	0.9
Cartosat-1, Warsaw	1A	2.5	1.4	0.8
OrbView-3, Zonguldak	1A	1.0 (2.0)	1.3	1.3
IKONOS, Zonguldak	2A	1.0	0.7	0.7
QuickBird, Zonguldak	2A	0.61	0.5	0.8
WorldView-1, Istanbul	2A	0.5	0.45	0.9
GeoEye-1, Riyadh	2A	0.5	0.4	0.8

projects listed in Table 6.4, there was a particular problem, in most cases, with limited accuracy of the control and check points. In the case of OrbView-3 the ground resolution of 1 m is reached by an over-sampling of neighbouring pixels of a staggered CCD line by 50%, which means that the projected pixel size is 2 m, avoiding higher object point accuracy. For KOMPSAT-1, only control points with a limited accuracy have been available; the image itself allows sub-pixel accuracy. With well-defined control points, sub-pixel accuracy is possible. Under operational conditions for the orientation of single scenes supported by control points, there are no problems reaching 1.0 GSD by a bias-corrected sensor-oriented RPC solution or geometric reconstruction. For mapping purposes, this is a satisfactory accuracy; the limiting factor is more the GSD itself, specifying the identification of objects.

6.5.3 Accuracy of blocks of images

Block adjustment of large numbers of space images is possible using only a limited number of control points. In most cases there is no stereoscopic coverage, so a block configuration such as for a 3D bundle block adjustment is not possible. Images can be joined using a modification of the old Anblock adjustment (Figure 6.37).

The Anblock adjustment is a combined 2D affine transformation of basic units. Without control points in the centre, a combined affine transformation has numerical problems. So instead of an affine transformation, a similarity transformation, or even

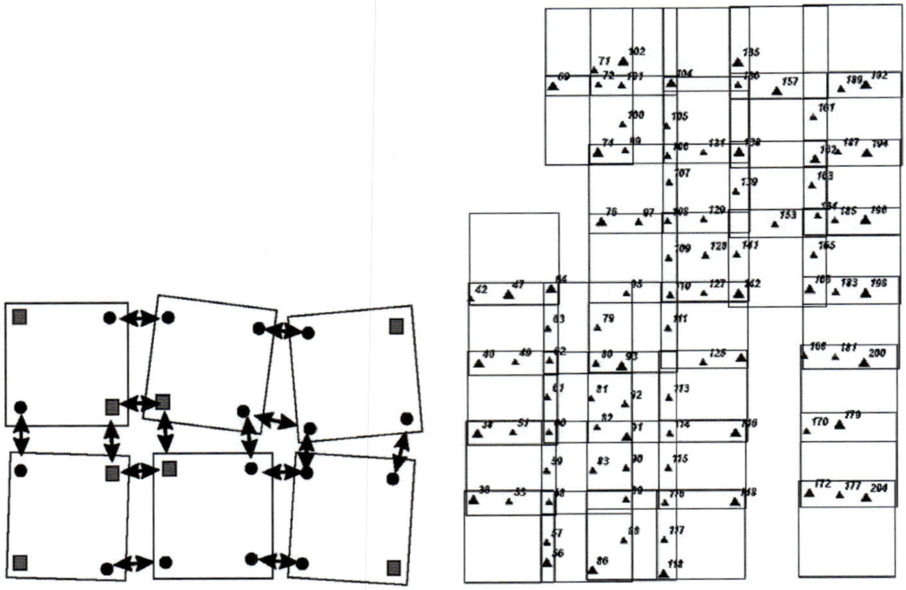

(a) (b)

Figure 6.37 (a) Principle of Anblock adjustment; and (b) block of QuickBird scenes.

shift-only parameters, may be used. Tie points have to be determined in overlapping areas, allowing a 2D tie of the images. Because of the basic image geometry, height values of the tie points are required; a solution to this can be obtained in an iterative manner with a digital elevation model. For most of the Earth, height values from the Shuttle Radar Topography Mission (SRTM, Section 7.4.2) are available; also ASTER GDEM (Section 7.4.3), which covers a larger area, can be used.

In open and flat areas, where tie points are determined, the SRTM height model may have a standard deviation of 3 to 6 m and ASTER GDEM 6 to 9 m, which is good enough for space images that do not have a large inclination. Figure 6.37(b) shows a block of 39 QuickBird images [Passini and Jacobsen, 2006], oriented with 5 to 89 control points. With 25 control points, an accuracy in the region of 3 GSD was reached. Grodecki and Dial (2003) report standard deviations of the horizontal coordinate components in the region of 1.0 GSD for a RPC block of IKONOS stereo scenes with just three ground control points, in this case not requiring a digital elevation model for support.

References

Baltsavias E. and Stallmann D., 1992. Metric information extraction from SPOT images and the role of polynomial mapping functions. *International Archives of Photogrammetry and Remote Sensing*, 29(B4), 358–364.

Chander, G., 2008. Bi-directional reflectance distribution function (BRDF) of natural surfaces. Civil Commercial Imagery Workshop (JACIE), Fairfax, VA, USA.

Dial, G. and Grodecki, J. (GeoEye), 2003. IKONOS stereo accuracy without ground control. ASPRS Annual Conference, Anchorage, AK, USA.

Dowman, I. and Dolloff, J., 2000. An evaluation of rational functions for photogrammetric restitution. *International Archives of Photogrammetry and Remote Sensing*, 29(B3), 254–266.

Dowman, I. and Michalis, P., 2003. Generic rigorous model for along track stereo satellite sensors. ISPRS Workshop High Resolution Mapping from Space, Hannover, Germany, http://www.ipi.uni-hannover.de/fileadmin/institut/pdf/dowman.pdf (accessed February 2011).

Fraser, C., Hanley, H. and Yamakawa, T., 2002. Three-dimensional geopositioning accuracy of Ikonos imagery. *The Photogrammetric Record*, 17(99), 464–479.

Fritsch, D. and Stallmann, D., 2000. Rigorous photogrammetric processing of high resolution satellite imagery. *International Archives of Photogrammetry and Remote Sensing*, 33(B1), 313–321.

Grodecki, J., 2001. IKONOS stereo feature extraction – RPC approach. ASPRS Annual Conference, St. Louis, MO, USA.

Grodecki, J. and Dial, G., 2003. Block adjustment of high resolution satellite images described by rational polynomials. *Photogrammetric Engineering and Remote Sensing*, 69(1), 59–68.

Gugan, D.J. and Dowman I.J, 1988. Accuracy and completeness of topographic mapping from SPOT Imagery. *The Photogrammetric Record*, 12(72), 787–796.

Gupta, A., Nain, J.S., Singh, S.K., Srinivasan, T.P., Gopala Krishna, B. and Srivastava, P.K., 2008. Long strip modelling for Cartosat-1 with minimum control. *International Archives of the Photogrammetry, Remote Sensing and Spatial Information Sciences*, 37(B1), 717–721.

Hanley, H.B. and Fraser, C.S., 2001. Geopositioning accuracy of Ikonos imagery: Indications from two dimensional transformations. *The Photogrammetric Record*, 17(98), 317–329.

Hanley, H.B., Yamakawa, T. and Fraser, C.S., 2002. Sensor orientation for high resolution imagery. *International Archives of the Photogrammetry, Remote Sensing and Spatial Information Sciences*, 34(B1), 69–75.

Hattori, S., Ono, T., Fraser, C. and Hasegawa, H., 2000. Orientation of high resolution satellite images based on affine projection. *International Archives of Photogrammetry and Remote Sensing*, 33(B3/1), 359–366.

Jacobsen, K., 1986. Geometric aspects of the use of space photographs. *ESA Proceedings of the International Symposium on Progress in Imaging Sensors*, Stuttgart, Germany, 427–432.

Jacobsen, K., 1994. Comparison of mapping with MOMS and SPOT images. *International Archives of Photogrammetry and Remote Sensing*, 30(B4), 225–230.

Jacobsen, K., 1997. Calibration of IRS-1C PAN camera. ISPRS Joint Workshop Sensors and Mapping from Space, Hannover, Germany, http://www.ipi.uni-hannover.de (accessed October 2008).

Jacobsen, K., 2002. Mapping with IKONOS images. 22nd EARSeL Symposium on Remote Sensing, Prague, Czech Republic. In Benes, T. (ed.), *Geoinformation for European-wide Integration*, Millpress, Rotterdam, Netherlands, 149–156.

Jacobsen, K., 2007. Digital Height Models by Cartosat-1, ISPRS Hannover Workshop, *International Archives of the Photogrammetry, Remote Sensing Spatial Information Sciences* (36 1/w 51).

Jacobsen, K., 2008. Geometric modelling of linear CCDs and panoramic imagers. In Li, Z., Chen, J. and Baltsavias E., *2008 ISPRS Congress Book*, CRC Press, London, UK, 145–155, http://www.ipi.uni-hannover.de/uploads/tx_tkpublikationen/III-2-jacobsen-1.pdf (accessed February 2011).

Jacobsen, K. and Passini, R., 2003. Comparison of QuickBird and IKONOS for the generation of orthoimages. ASPRS Annual Conference, Anchorage, AK, USA.

Kocaman, S. and Gruen, A., 2007. Orientation and calibration of ALOS/PRISM imagery. *International Archives of the Photogrammetry, Remote Sensing and Spatial Information Sciences*, 36(I/W5), http://www.isprs.org/proceedings/XXXVI/1-W51/paper/Kocaman_Gruen.pdf (accessed September 2011).

Konecny, G., Lohmann, P., Engel, H. and Kruck, E., 1987. Evaluation of SPOT imagery on analytical photogrammetric instruments. *Photogrammetric Engineering and Remote Sensing*, 53(9), 1223–1230.

Kratky, V., 1989. Rigorous photogrammetric processing of SPOT images at CCM Canada. *ISPRS Journal of Photogrammetry and Remote Sensing*, 44(2), 53–71.

NGA, 2011. http://earth-info.nga.mil/GandG/publications/tr8350.2/wgs84fin.pdf (accessed March 2011).

Okamoto A., 1981. Orientation and construction of models: Part III Mathematical basis of the orientation problem of one-dimensional central perspective photographs. *Photogrammetric Engineering and Remote Sensing*, 47(12), 1739–1752.

OrbImage, 2004. OrbView-3 geometric calibration and geopositional accuracy. JACIE High Spatial Resolution Commercial Imagery Workshop, http://calval.cr.usgs.gov/JACIE_files/JACIE04/files/1Mulaw15.pdf (accessed September 2011).

Passini, R. and Jacobsen, K., 2006. Accuracy analysis on large blocks of high resolution images. *International Archives of the Photogrammetry, Remote Sensing and Spatial Information Sciences*, 36A (PS1-16).

Schroeder, M., Müller, P. and Reinartz, P., 1999. Vicarious radiometric calibration of MOMS at La Crau test site and intercalibration with SPOT. Workshop on Sensors and Mapping from Space, Hannover, Germany.

Seo, D.C., Yang, J.Y., Lee, D.H., Song, J.H. and Lim, H.S., 2008. KOMPSAT-2 direct sensor modelling and geometric calibration/validation. *International Archives of the Photogrammetry, Remote Sensing and Spatial Information Sciences*, 37(B1/WG1/1), 47–52.

Smiley, B. (DigitalGlobe), 2008. The absolute and relative geolocation accuracies of QB02 and WV01. JACIE Civil Commercial Imagery Workshop, Fairfax, VA, USA.

Srinivasan, T.P., Islam, B., Sanjay, K., Singh, K., Gopapa Krishna, B. and Srivatava, P.K., 2008. In-flight geometric calibration – an experience with Cartosat-1 and Cartosat-2. *International Archives of the Photogrammetry, Remote Sensing and Spatial Information Sciences*, 37(B1/WG1/1), 83–88.

Tansock, J., Thurgood, A. and Larsen, M., 2004. System-level approach to characterization and radiometric calibration of a space-based electro-optical sensor. In Morain, S. and Budge, A. (Eds.), *Post-Launch Calibration of Satellite Sensors*, Taylor Francis Group, London, UK, 59–65.

Toutin, Th., 2000. Elevation modeling from satellite data. In Meyers, R.A. (Ed.), *Encyclopedia of Analytical Chemistry: Applications, Theory and Instrumentation*, Vol. 10, John Wiley & Sons, Ltd., Chichester, UK, 8543–8572.

Toutin, Th., 2004. Comparison of stereo-extracted DTM from different high resolution sensors: SPOT-5, EROS-a, IKONOS II, and QuickBird. *IEEE Transactions on Geoscience and Remote Sensing*, 42(10), 2121–2129.

Toutin, Th. and Cheng, P., 2000. Demystification of IKONOS. *Earth Observation Magazine*, 9(7), 17–21.

Valorge, C., Meygret, A., Lebègue, L., Henry, P., Bouillon, A., Gachet, R., Breton, E., Léger, D. and Viallefont, F., 2004. Forty years of experience with SPOT in-flight calibration. In Morain, S. and Budge, A. (Eds.), *Post-Launch Calibration of Satellite Sensors*, Taylor Francis Group, London, UK, 119–133.

Weser, T., Rottensteiner, F., Willneff, J., Poon, J. and Fraser, C., 2008. Development and testing of a generic sensor model for pushbroom satellite imagery. *The Photogrammetric Record*, 23(123), 255–274.

7 Processing and Products

Karsten Jacobsen

7.1 Introduction to matching

In order to obtain 3D information about an object, it must be visible in at least two images taken from different positions or in one image together with a height model. After image orientation, object positions can be measured manually; this is the dominant factor in the information acquisition for mapping because automatic object recognition required for topographic mapping is complex and today only the first steps have been taken towards automation. For digital elevation models this is different: they require a high number of points, and only corresponding image points without attribute information have to be determined. Nevertheless automatic image matching for obtaining homologue points is not a simple task and still has some problems and limitations, so usually post-processing by filtering and/or manual editing has to be carried out.

In optical as well as radar images the visible surface is presented in the image, so digital elevation models based on such images are digital surface models (DSMs) with points on the visible surface, while in topographic maps digital terrain models (DTMs) with points on the bare ground are used (Figure 7.1). The expression "digital elevation model" (DEM) is a generic term used for both DSMs and DTMs. The term "digital height model" is also used. If a DSM includes enough points on the bare ground, which are well distributed, it can be filtered to a DTM [Passini *et al.*, 2002]; this can be successful in open areas.

Automatic image matching requires similarity of objects in images. If the images used for matching are not taken within a short time interval, the object itself and the illumination may change, causing problems for matching. With higher resolution such problems increase because of lower integration of details in one pixel. With the exception of the limited applications using space photographs, such as from the metric camera and the large format camera, the civilian use of space images started with SPOT-1 in 1986. SPOT satellites are able to take stereo image combinations by

Figure 7.1 (a) DSM with points on top of the visible surface; and (b) DTM with points only on the bare ground.

Figure 7.2 Stereo pair of SPOT images taken in June and August of the same year.

viewing to the side (Figure 4.10), as is also the case with IRS-1C/1D and KOMPSAT-1. So the second image of a stereo combination can be taken, at the earliest, one day later. If on the next day clouds do not allow imaging, there may be a time interval of several days or longer, which may lead to problems for image matching. This can be seen in the example of the image pair shown in Figure 7.2, where the land cover partially changed from green in June to yellow in August. Today, with flexible satellites such as IKONOS, QuickBird, EROS, Cartosat-2, WorldView-1 and 2 and GeoEye-1, the satellites can rotate quickly, allowing stereoscopic data acquisition from the same orbit (Figure 4.1), with a difference in time in the region of one minute. Stereoscopic coverage of larger areas is possible with stereo satellites such as ASTER, Cartosat-1, ALOS PRISM and SPOT-5 HRS, having two or three optics (Figures 4.18, 4.23, 4.20 and 4.16) for large area stereoscopic imaging.

7.2 Preparation of images

Automatic image matching is a 2D problem; corresponding image points may have differences in x and y image coordinates. Based on known image orientation, the effort of searching for corresponding points in a stereo pair can be reduced to a search in one

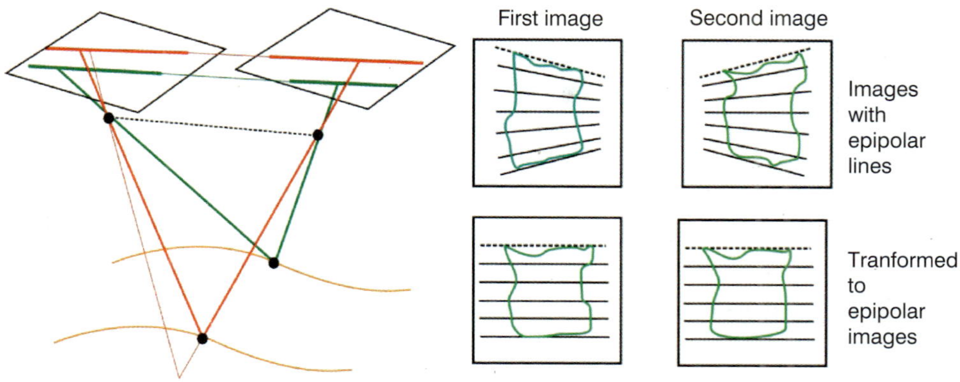

Figure 7.3 Epipolar lines and transformation images from epipolar lines to epipolar images.

direction by geometric constraints or by using epipolar images. In epipolar images the plane defined by an object point and both projection centres is intersected with lines having the same *y* coordinates on both images. Perspective images of a stereo pair which are not epipolar can be transformed to epipolar images (Figure 7.3). The transformation is simple for perspective images, but line-scan images also can be transformed to epipolar images using an iterative process with sufficient accuracy. Level 2A type line-scan images (projected to a plane of constant height) can simply be transformed into epipolar images by rotating the images to the base direction of the stereo model. The *y*-parallaxes of these quasi-epipolar space images depend on the height differences with respect to the reference height level, the influence of the Earth's curvature and the corresponding angle at the actual projection centre. Under normal conditions the *y*-parallaxes do not exceed 0.1 GSD [Alobeid *et al.*, 2011]. As mentioned, in epipolar images corresponding points have the same *y* coordinates, so the search for corresponding points need only be carried out in the *x* direction. Epipolar images are limited to a combination of two images. If more images are to be used for matching, a geometric constraint has to be used instead of epipolar images.

A general contrast enhancement of images before automatic image matching has no positive effect, because usual matching methods are based on normalised sub-images. Also, most filter methods do not improve the matching. This may be different for local filters, for example those for de-striping or using a Wallis filter with a small window size.

Image artefacts disturb automatic image matching and should be removed as far in advance as possible. A typical problem is striping as in ASTER, ALOS/PRISM and EROS images. In ALOS/PRISM images, JPEG artefacts (Section 3.6.4.1) can be seen as blocking. Striping and blocking can be reduced or eliminated by using a low-pass filter, but this causes loss of information, which can be avoided by filters optimised for the individual image problem. Striping of ASTER images is caused by the staggered CCD lines having shifts of the grey values with respect to each other. A simple and effective correction for ASTER only shifts the grey values by half the difference of the grey value averages for the even and odd columns as shown in Figure 7.4.

The de-striping of EROS-B images is more complex than the correction of ASTER images. The grey value differences of neighbouring columns are dependent on the grey value itself, and even with a complex correction of the grey values no satisfactory results have been achieved by filtering in the image domain. This is different for filtering in the frequency domain. After elimination of disturbing parts in the frequency domain, by inverse Fourier transform of the modified power spectrum, an image less affected by striping can be generated (Figure 7.5). Other artefacts such as saturation, typical for several space sensors, do not play an important role, because they appear in only one image and lead to voids in the DSM, but not to blunders in height.

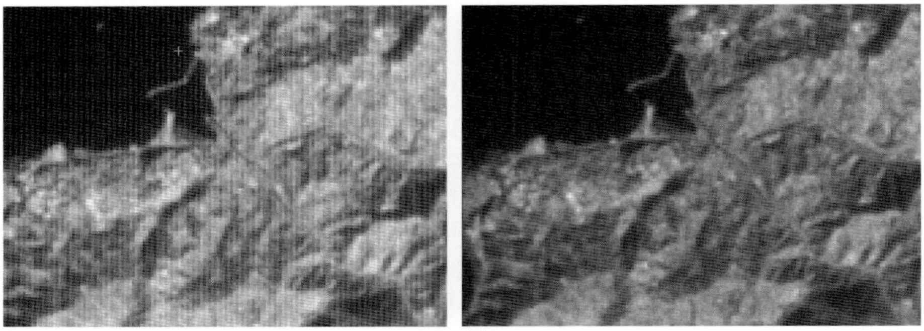

(a) (b)

Figure 7.4 ASTER sub-image disturbed by striping: (a) before correction; and (b) after correction.

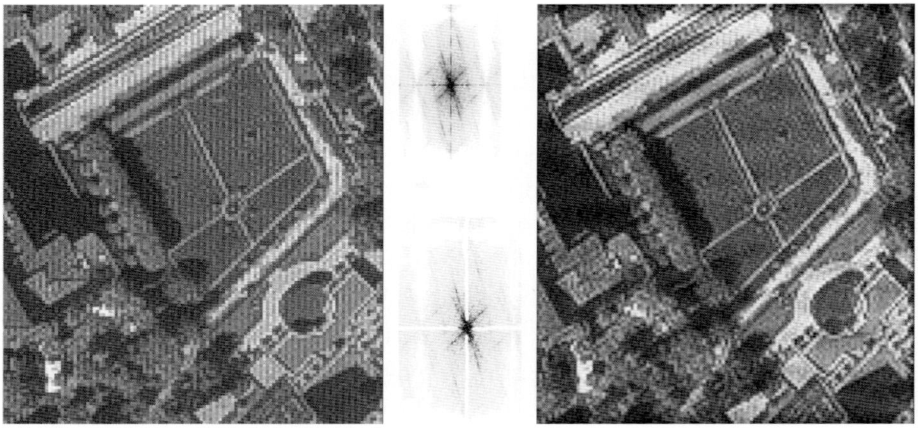

(a) (b) (c)

Figure 7.5 (a) EROS-B sub-image showing striping; (b) power spectrum of the image (top) and modified power spectrum (bottom); and (c) destriped image after using an inverse Fourier transform of the modified power spectrum.

The contrast of an image can be enhanced with a Wallis filter, which is good for the human operator, but for matching it only has an advantage if the filter matrix is small. The Wallis filter is a local statistical filter. For the specified window around the processed pixel with the grey value G_{in}, the mean grey value, M_{in}, and the standard deviation, S_{in}, of the grey values are computed. The grey value of the pixel will be changed to G_{out} based on an average grey value (desired mean M_{out}) and a chosen standard deviation (desired standard deviation S_{out}). In addition, the relationship S_{out}/S_{in} can be limited to a chosen value. The Wallis filter enhances the local contrast, so more details in both the dark and the bright parts of the original image can be seen. G_{out} is calculated from the equation

$$G_{out} = M_{out} + (G_{in} - M_{out}).\frac{S_{out}}{S_{in}} \tag{7.1}$$

It is clear from Figure 7.6 that the local contrast is improved by using a Wallis filter with a small window; the filter image shows significantly more details in dark shadows as well as in bright areas. Such an improvement also supports automatic image matching. Depending on the images, adaptive noise reduction, sharpening of edges and preservation of fine details can be performed and may improve image matching [Baltsavias *et al.*, 2001].

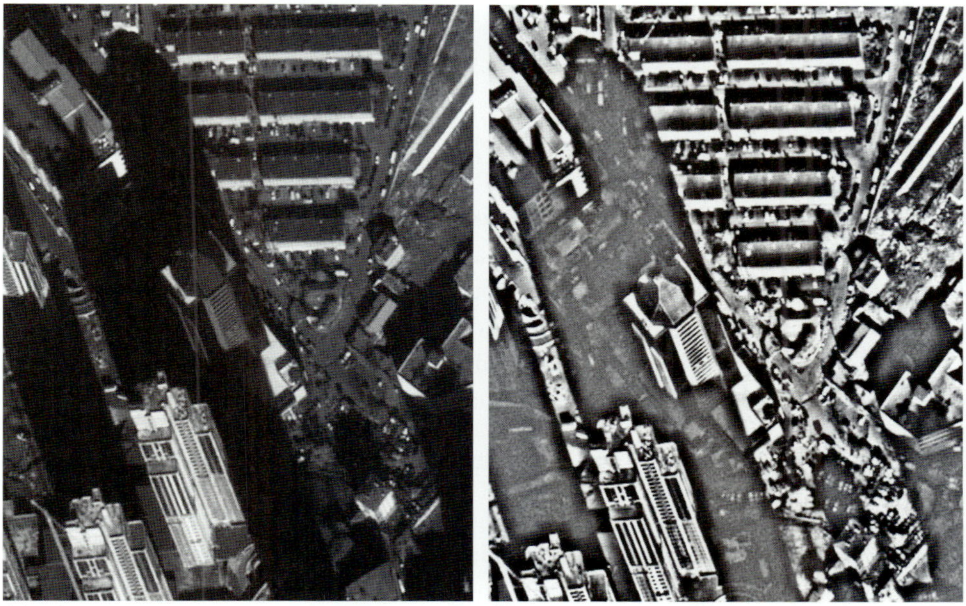

(a) (b)

Figure 7.6 (a) Original WorldView-1 sub-image; and (b) same image improved using a local Wallis filter.

7.3 Generation of digital elevation models

7.3.1 Image matching

7.3.1.1 Introduction

The generation of a DEM, or digital height model, is one of the key issues for optical space images. The accuracy and details, mainly controlled by the DSM point spacing, are dependent on the GSD of the sensor and the imaging configuration, dominated by the base-to-height ratio (Section 1.3.4). Because of the clear economic advantages of automatic image matching, no manual measurement of height models is normally done. The height of the visible surface (DSM) is determined by automatic image matching; the precision of the height determination is given by

$$\sigma_Z = \frac{H}{B} \cdot \sigma_{px} \tag{7.2}$$

where σ_Z is the standard deviation of the object height; H is the height above ground; B is the image base (distance of projection centres); and σ_{px} is the standard deviation of the x-parallax.

Equation (7.2) is a general equation for the standard deviation of the object height and depends on the standard deviation of the x-parallax. In the case of digital images, σ_{px} can be expressed as a fraction of the GSD, leading to the equation

$$\sigma_Z = \frac{H}{B} \cdot a \cdot GSD \tag{7.3}$$

where σ_Z is the standard deviation of the object height based on digital images, and a is the fraction of ground sampling distance (GSD). The factor a in Equation (7.3) depends on the land cover of the object area (for example open, forest or urban), the contrast of the area, the matching method and the base-to-height ratio. With a small angle of convergence, the images are more alike, reducing the factor a, so the accuracy depends to a lesser extent on the base-to-height ratio.

Image matching may be area based, feature based or a combination of both of these; or it may be pixel based with an additional cost function.

7.3.1.2 Cross-correlation

The traditional method of image matching began with area-based cross-correlation. Image matching by cross-correlation is based on the geometric relationship of how a sub-image of one image fits to the sub-image in the other image (Figure 7.7). That means the image planes or the transformed image planes, as well as the object plane, must be parallel, must have the same scale and should not be rotated with respect to each other, while the vertical line locus (VLL) transforms a sub-matrix of the object plane (which is expected to be horizontal) into the images. In both cases the similarity of the image sub-matrixes is computed by cross-correlation.

The formula for the coefficient of correlation r is given by

$$r = \frac{\sum_{i=1}^{n}(g_{a_i} \cdot g_{b_i}) - \frac{1}{n}(\sum_{i=1}^{n} g_{a_i}) \cdot (\sum_{i=1}^{n} g_{b_i})}{\sqrt{(\sum_{i=1}^{n} g_{a_i}{}^2 - \frac{1}{n}(\sum_{i=1}^{n} g_{a_i})^2)(\sum_{i=1}^{n} g_{b_i}{}^2 - \frac{1}{n}(\sum_{i=1}^{n} g_{b_i})^2)}} \tag{7.4}$$

where $-1 \le r \le +1$. The quantity r is a normalised value expressing the relative similarity of the sub-matrixes in the used images. The pattern matrix is shifted over the search matrix until the highest correlation coefficient is found (Figure 7.8). The standard

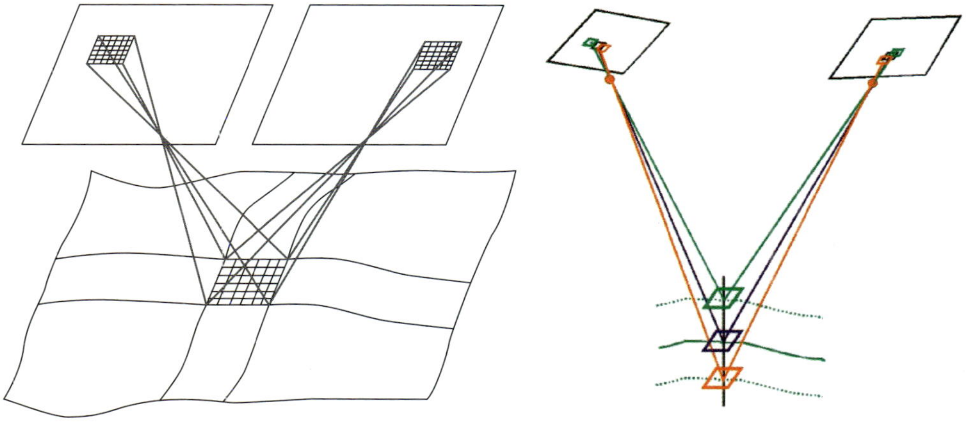

(a) (b)

Figure 7.7 (a) Sub-matrixes in two images used for cross-correlation; and (b) vertical line locus.

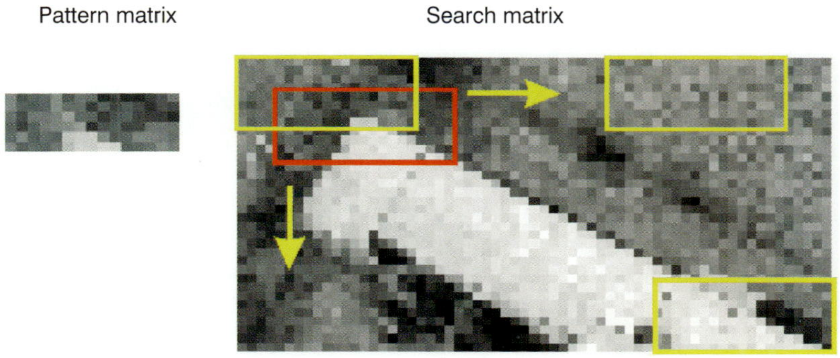

Figure 7.8 The process of cross-correlation with the pattern matrix shifted, pixel by pixel, over the search matrix in the x and y directions. The largest correlation coefficient matching the pattern matrix with the search matrix is shown in red.

cross-correlation does not use the image orientation and has to search in the x and in y direction, while the VLL, using the exterior orientation, only has to search in the direction of the epipolar line. The VLL has the advantage that the object coordinates X and Y are fixed, directly allowing the height determination in a raster of object points, without any interpolation in the object space. In addition, for the VLL, the images may be rotated – only the object is treated as a plane within the handled matrix size.

7.3.1.3 Least-squares matching

Least-squares matching has proved to be very powerful because it has no restriction introduced by the scene orientation and inclination of the terrain. The basic equations are set up in the context of a statistical estimation model

$$\begin{aligned} x'' &= a_0 + a_1 . x' + a_2 . y' \\ y'' &= b_0 + b_1 . x' + b_2 . y' \\ g'(x', y') &= r_0 + r_1 . g''(x'', y'') \end{aligned} \tag{7.5}$$

where x' and y' are coordinates on one image; x'' and y'' are coordinates on the other image; and g' and g'' are first and second differentials of the grey level values on the two images. A least-squares solution is applied to solve for the unknowns a_0–a_2, b_0–b_2 and r_0–r_1. With least-squares matching, the image orientation may be random. It is assumed that there is a constant terrain slope within the search window (Figure 7.9). Constant and linear grey value differences are taken into account.

The window used for least-squares matching should not be too small to ensure a satisfactory signal power, but neither should it be too large, causing a smoothing effect.

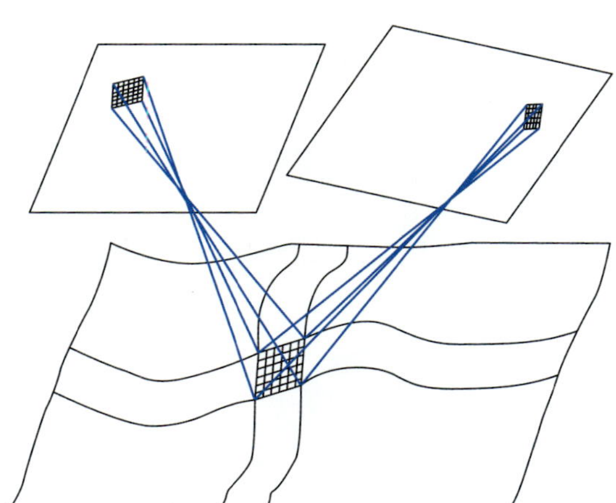

Figure 7.9 Matching of patches from two images using least-squares estimation.

By adaptive least-squares matching the optimal window size can be identified by the matching process. The least-squares matching has limited search radius, requiring good approximations, which may come, for example, from cross-correlation, feature-based matching or region growing. On the other hand, the approximations for cross-correlation and feature-based matching are not as sensitive and may be solved by image pyramids or region growing. Nevertheless, all mentioned methods of image matching require some approximations.

7.3.1.4 Feature-based matching

Feature-based matching is usually based on "interest operators", which are algorithms that detect features of interest in an image, such as corners, edges or regions. The classical Foerstner operator has been widely adopted

$$N=\begin{bmatrix}\sum g_x'^2 & \sum g_x' g_y' \\ \sum g_x' g_y' & \sum g_y'^2\end{bmatrix}$$

$$Q=N^{-1}=\frac{1}{\det(N)}\begin{bmatrix}\sum g_y'^2 & \sum g_x' g_y' \\ \sum g_x' g_y' & \sum g_x'^2\end{bmatrix} \tag{7.6}$$

$$q=1-\left(\frac{\lambda_1-\lambda_2}{\lambda_1+\lambda_2}\right)^2=\frac{4\det(\mathrm{N})}{\mathrm{spur}(\mathrm{N})^2}$$

where g is grey value; g'_x and g'_y are differential grey values in x and y directions; first criteria, w = spur(Q) shall be small; and second criteria, feature shall be round, with q values usually > 0.8. Use of the operator enables the location of distinct points as corners and centres of circular image features to be detected with sub-pixel accuracy as shown in Figure 7.10. The density of feature points is not normally sufficient to generate

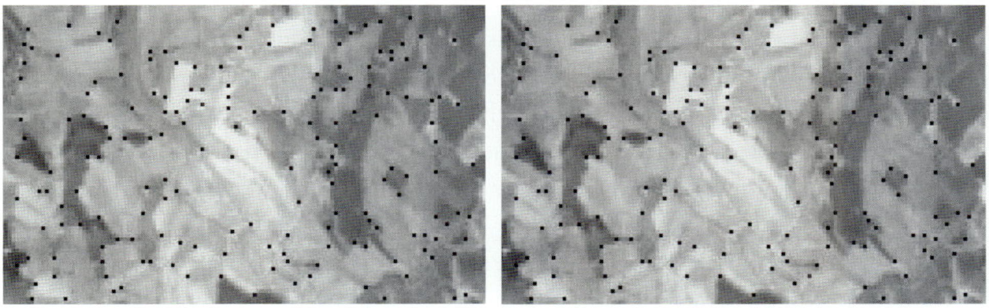

(a) (b)

Figure 7.10 Images of a stereo model with overlaid feature points.

a dense digital elevation model, but the feature points can be used as a preliminary rough estimation of the DSM and as initial information for area-based matching.

7.3.1.5 Combined methods

The standard area-based matching is not able to handle objects with sudden changes in height as in the case of buildings. With a combination of eccentric pattern matrixes, the fit to buildings can be improved using a combination of feature-based and area-based matching.

7.3.1.6 Pixel-based matching with dynamic programming

The real shape of buildings can be determined by pixel-based matching. This is not possible with single pixels, but can be done with cost functions respecting the neighbourhood, using for example dynamic programming [Birchfield and Tomasi, 1999] or semi-global matching [Hirschmüller, 2008]. For practical handling, both methods are applied to epipolar images, reducing the matching to a search in just one direction.

Pixel-based matching with dynamic programming (DP) is able to reduce errors at regions of sudden height changes, for example at building outlines, where area-based matching is known to perform poorly. No windows are required for matching; intensity values of individual pixels are compared, in combination with constraints to reward successful matches and to penalise occlusions. The algorithm focuses especially on generating correct results at height discontinuities, while sacrificing some accuracy in smooth areas.

Matching is formulated as an optimisation problem for each corresponding epipolar line pair, based on a pre-defined cost function to be minimised. Each pixel in the left epipolar line is compared to all pixels of the conjugate epipolar line, and a 2D array of costs is constructed. The cost function used, $\lambda(x,y)$, has three components

$$\lambda(x, y) = \sum_{i=1}^{N_m} d(x_i y_i) - N_m \,.\, K_r + N_{occ} \,.\, K_{occ} \tag{7.7}$$

The first component is the sum of the dissimilarities, $d(x_i,y_i)$, between the matched pixels; it should dominate the cost function. The second component ($N_m \times K_r$) is a reward for correct matching, where N_m is the number of matched pixels and K_r is the match reward per pixel. The third component ($N_{occ} \times K_{occ}$) is a penalty for occlusions, where N_{occ} is the number of occlusions (*not* the number of occluded pixels) and K_{occ} is the occlusion penalty. The factors K_r and K_{occ} are determined empirically.

The easiest dissimilarity function is the absolute value of differences in intensities. Instead, the dissimilarities are computed by using linearly interpolated intensities halfway between the pixels in each corresponding epipolar line and its neighbours, according to Birchfield and Tomasi (1998), to overcome sampling effects.

In dynamic programming, the shape of buildings can be identified without smoothing effects, as happens with area-based matching, but there is the disadvantage of handling neighbouring lines independently, causing some streaking of the building shape. This can be reduced by filtering across the line direction [Alobeid *et al.*, 2009].

A smoothness constraint within the global cost function for connecting the disparities of several line pairs in different directions, intersecting in one pixel simultaneously, has been introduced in semi-global matching (SGM) [Hirschmüller, 2008], which is also based on dynamic programming. There are two ideas involved in this matching method. First, the dissimilarity is expressed pixel by pixel by mutual information (MI). MI measures correspondence without assuming that conjugate points have similar intensity values. Instead, the joint probability distribution in the form of the joint intensity value histogram is used. MI has been shown to be rather robust with respect to radiometric differences [Hirschmüller and Scharstein, 2009]. The first step for SGM is to obtain an initial disparity image that is required for warping one of the stereo images before MI can be calculated. It starts with a random disparity image, and then continues in a hierarchical fashion. Subsequently, the joint histogram is derived over the whole images; it is stored as a 256 × 256 histogram and smoothed using a Gaussian kernel. Then, the MI values are computed. The second idea of SGM is a global 2D smoothness constraint across multiple intersecting lines at the pixel under consideration, by combining the various 1D paths towards this pixel derived from dynamic programming.

As is obvious from Figures 7.11 and 7.12, least-squares matching does not generate the correct building shape. Using dynamic programming, the building shape becomes clear, but it is disturbed by streaking, requiring a Median filter across the line direction. However, small details are lost with any filter. Using SGM, the building shapes are determined correctly.

7.3.1.7 Shape from shading

In the case of weakly textured images, the differential height of homogenous objects with grey levels close to Lambertian reflection can be determined by shape from shading

(a) (b) (c)

Figure 7.11 Profile of buildings after image matching with IKONOS scenes: (a) least-squares matching; (b) dynamic programming; and (c) semi-global matching [Alobeid *et al.*, 2009].

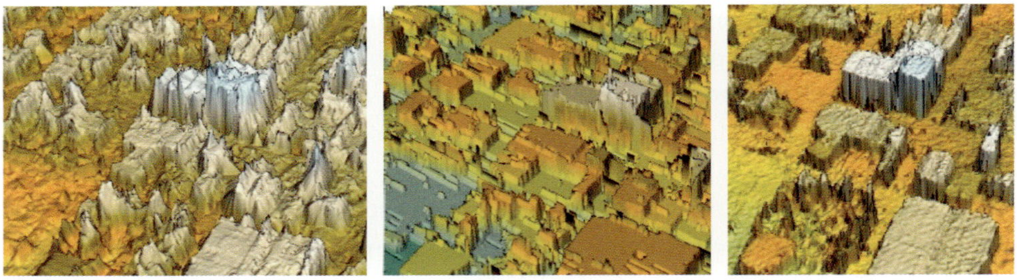

(a) (b) (c)

Figure 7.12 Oblique view of buildings after image matching with IKONOS scenes: (a) least-squares matching; (b) dynamic programming; and (c) semi-global matching [Alobeid *et al*., 2009].

[Heipke *et al*.. 2001], even with just one image. This method may be successful for extraterrestrial areas, as on the moon and Mars, and also for deserts. It cannot be used in areas with vegetation, which cause a variation of grey values, but contrast is available in such areas, enabling the use of standard methods.

7.3.1.8 Multiple image matching

Object heights can be determined with two images taken from different directions, corresponding to four observations and three unknowns. The over-determination is the *y*-parallax, but this does not allow checking the object height. So the object height cannot be controlled with just two images. Based on the assumption that the surface does not include sudden height changes, the object heights can be checked by means of neighbouring points. This assumption can be used in open areas, but not in cities. A true check is only possible with at least one more image. Blunders can be identified by multiple image matching; in addition, the accuracy can be improved by up to 30% as shown by Zhang and Gruen (2006). Figure 7.13 shows the geometry for a three-line sensor.

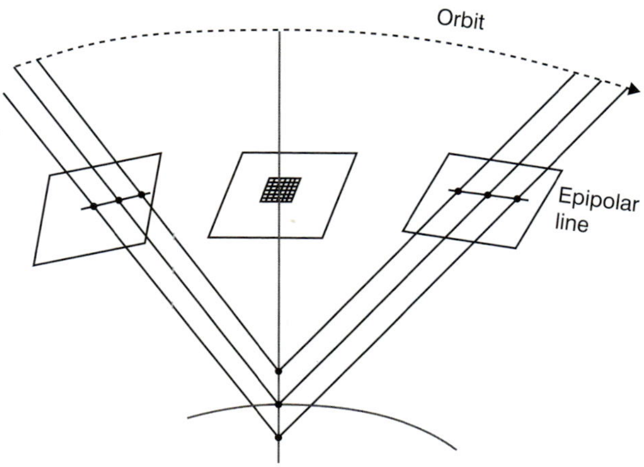

Figure 7.13 Geometric relationship of multiple image matching using image orientation.

Multiple image matching is limited by the fact that only a restricted number of three-fold scenes are available, with the exception of images from ALOS PRISM equipped with three cameras (Section 4.3.5).

7.3.2 Analysis of height models

The output from image matching is normally a point cloud with an irregular distribution. However, digital elevation models are frequently arranged in a regular raster as shown in Figure 7.14. Except for points determined by a vertical line locus, the height values of the raster arrangement are interpolated from points in the original point cloud. So the digital height model is influenced by the accuracy of the original height points as well as the interpolation. The interpolation has a stronger influence in rough terrain than in mountainous areas.

The most important factor determining the accuracy of a digital elevation models is the GSD of the images used. This may not be identical to the size of the projected pixel, as in the case of staggered CCD lines. According to Equation (7.3), the vertical accuracy depends on the GSD, the base-to-height ratio and an empirical factor. Also the point spacing of the generated height model is dominated by the GSD. In the case of point spacing of just 1 GSD and area-based matching, neighbouring height values are strongly correlated. For the correct height information in a stereo model, a point spacing of 2 to 3 GSD is more realistic. The point spacing is important for the morphologic details included in the DSM (Figure 7.15).

A complete accuracy evaluation of digital height models is only possible with precise reference height models. In addition to the standard deviation of the height (σ_Z), the horizontal fit should also be evaluated. A typical limitation of height models is caused by datum errors of the coordinate reference and orientation problems of the stereo models used (Figure 7.16), requiring a check of the investigated height model

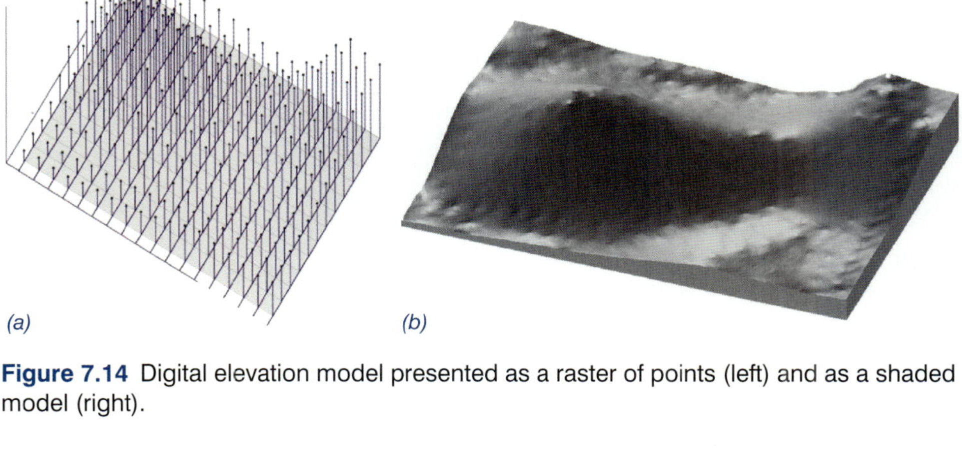

(a) (b)

Figure 7.14 Digital elevation model presented as a raster of points (left) and as a shaded model (right).

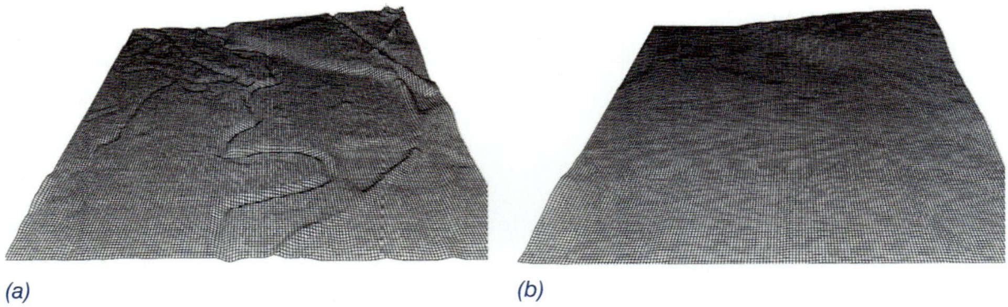

Figure 7.15 Digital height models of the same area with different point spacings: (a) 12.5 m spacing; and (b) 50 m point spacings.

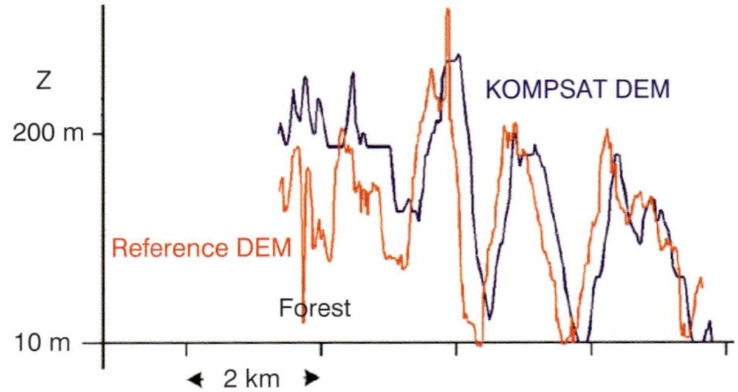

Figure 7.16 Shift of height profiles of KOMPSAT-1 DSM with respect to the reference DTM.

for shifts in all three coordinate components. As shown in Figure 7.16, the investigated KOMPSAT-1 DSM is shifted approximately 200 m in the profile direction. On the left-hand side it is influenced by the height of a forest. For a correct height analysis, a horizontal shift by adjustment to the reference height model has to be done in advance.

A complete and correct analysis cannot be done with individual check points. Check points are usually located in areas with good contrast, undisturbed by vegetation and height changes within the window of interpolation used, leading to results which are better than average. The orientation can be investigated with check points, but for digital height models the accuracy in most cases is overestimated by a factor of two or even more.

With automatic image matching using optical space images, digital surface models are generated; the heights of vegetation and buildings lead to differences between the DSM and a reference DTM. For mapping purposes, a DTM is requested because of changes in height of vegetation. The accuracy analysis should be made separately for any terrain type. If the terrain includes enough areas with height values of the bare ground, the influence of vegetation and buildings can be eliminated, or at least reduced, by filtering [Passini *et al.*, 2002]. The accuracy evaluation should take into account the terrain inclination because nearly all height models show a dependency on the tangent of terrain inclination, requiring an expression of the accuracy, such as $\sigma_Z = a + b \cdot \tan\alpha$, where σ_Z is the standard deviation of the height, α is the terrain inclination (see also Figure 7.17) and a and b are the characteristics to be determined.

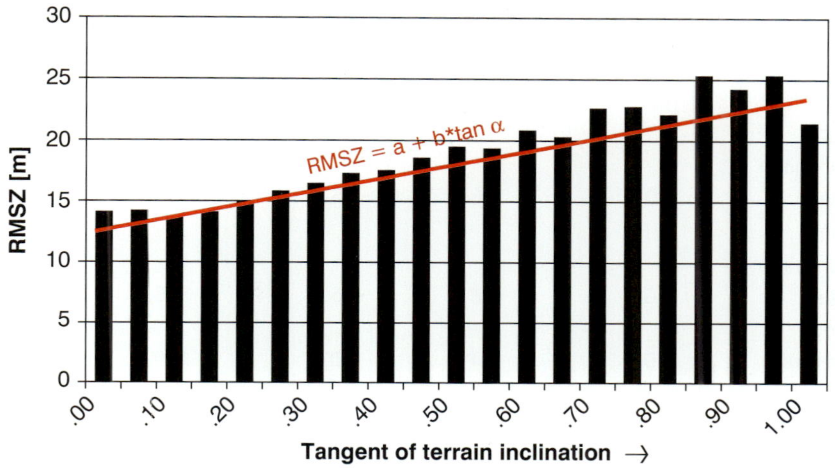

Figure 7.17 Standard deviation of an ASTER DEM, separately computed for different terrain inclination classes.

The influence of the interpolation depends on the roughness of the terrain. The roughness can be expressed by the difference of the neighbouring height differences in a profile:

$$roughness = \sqrt{\frac{\sum_{i=1}^{i=n} (\Delta h_i - \Delta h_{i-1})^2}{n}} \tag{7.8}$$

The roughness depends approximately on the square of the spacing. So if in the rough mountainous area of Zonguldak the loss of accuracy by interpolation corresponds to 12 m (Table 7.1) for a DTM with 80 m spacing, it can be estimated as 3 m for a point spacing of 40 m. On average, the influence on the interpolation is approximately 50% of the roughness (Table 7.1) [Jacobsen, 2005a, 2005b]. The influence on the interpolation can be estimated to be one-eighth of the roughness computed with the differential height differences of the investigated height model. Corresponding to this relationship, by standard error propagation the accuracy of the height model can be estimated to be

$$\sigma_Z = \sqrt{component\ of\ roughness^2 + S^2{}_Z} \tag{7.9}$$

Table 7.1 Terrain characteristics and rmse loss of accuracy by interpolation [Jacobsen 2005b]

	Spacing (m)	Average terrain inclination (°)	Average change of terrain inclination (°)	Roughness (m)	$rmse_z$ of interpolation (m)
Zonguldak, rough mountain	80	15.0	18.0	25.6	12.0
Arizona, smooth mountain	90	9.6	5.0	8.1	4.8
Bavaria, rolling	80	4.8	6.8	9.6	2.1
New Jersey, flat	60	1.4	0.9	0.9	0.45
New Jersey, flat	120	1.4	0.9	1.8	1.12

where σ_Z is the standard deviation of the height point directly determined by image matching.

The dependency of the height accuracy on aspect (the direction of terrain inclination) will not be the same in all directions. This characteristic is typical for a DSM from interferometric synthetic radar as shown in Figure 7.18(a) [Passini and Jacobsen, 2007b], but may also be evident for optical data as shown by an example of SPOT HRS in the Alps (Figure 7.18b). In the test area Inzell, the slope-dependent component B of the accuracy relationship $rmse_Z = A + B \,.\, \tan \alpha$ is larger in the case of terrain inclination in a north-east direction.

An example of the effect of DSM filtering can be seen in Figure 7.19. The original DSM, derived from SPOT HRS images, shows the influence of the forest, while the filtered DSM is more similar to the reference DTM [Jacobsen, 2003]. With the exception of dense forest, the accuracy of height models is improved by filtering. Blunders can be detected and eliminated if they are large enough in relation to height differences of the surface.

According to Equation (7.3), the standard deviation of the height, identical to LE68 (linear error with 68% probability), depends on the base-to-height ratio $B{:}H$. The most usual $B{:}H$ is in the region of 0.6, corresponding to an angle of convergence

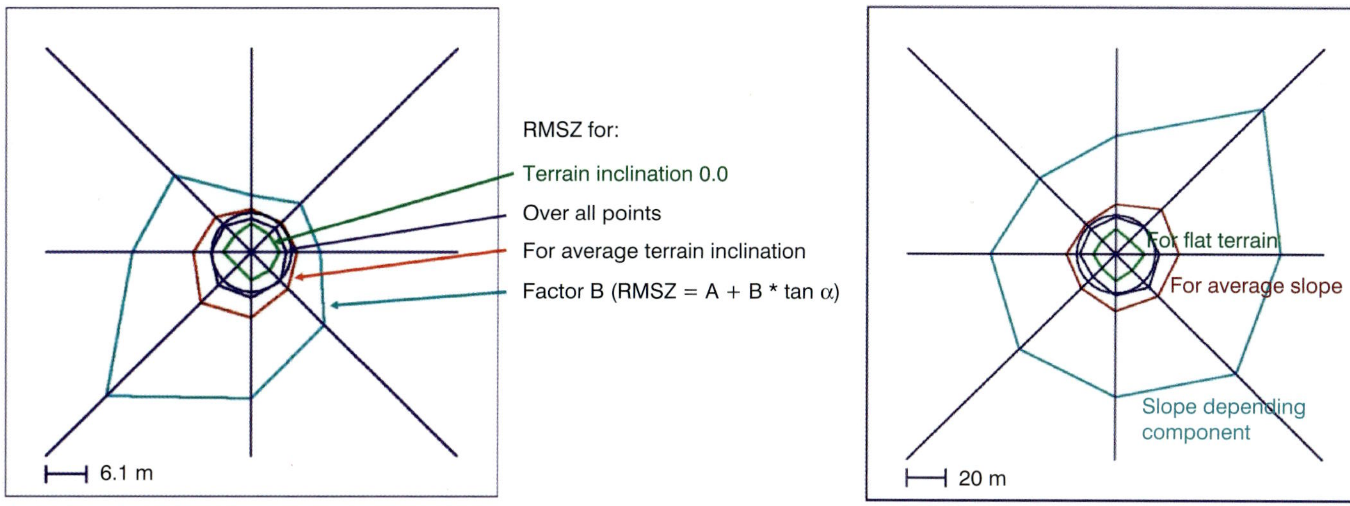

Figure 7.18 Standard deviation of height as a function of aspect (direction of terrain inclination): (a) SRTM height model of Zonguldak; and (b) SPOT HRS of Inzell.

(a) (b) (c)

Figure 7.19 Influence of a DSM filter: (a) original DSM from SPOT HRS; (b) filtered DSM; and (c) reference DTM.

of 35°. A larger angle of convergence causes larger differences of the images in a stereo model, which does not lead to improved results for automatic image matching. A smaller angle of convergence, or smaller *B*:*H* value, results in the images of the stereo model being more similar, improving the matching accuracy, and this may at least partially compensate for the geometric influence of the angle of convergence.

Over bare surfaces, the standard deviation of the height is not far away from 1.0 GSD, but it is becoming more difficult to achieve for very high resolution images [Toutin, 2005]. Multiple image matching slightly improves the object accuracy; nevertheless, the dominating effect of DEM generation with IKONOS images is the object definition. On bare ground areas, with multiple image matching, Zhang and Gruen (2006) reached a standard deviation of the height between 1.15 m and 2.14 m. For city areas and trees/bushes, the value obtained for σ_Z between 2.8 m and 8.2 m is dominated by the object definition.

An overview of the DSM accuracy based on different satellite images is given in Table 7.2 [Jacobsen, 2005a, 2005b, 2006]. It confirms the results mentioned before. The accuracy of the DSMs based on different sensors can be compared using Equation (7.2) with the standard deviation of the x-parallax ($rmse_{px}$) in GSD units. For open and flat areas $rmse_{px}$ in most cases is slightly below 1 GSD. This can be improved by filtering. Against check points the accuracy is unrealistically small because of the above mentioned special conditions of check points. A time delay between the images reduces the accuracy, reliability and completeness of the DSMs based on automatic image matching. Only in the case of DSMs in areas with little or no vegetation, as in Arizona, a limited time delay does not cause problems. In forest areas the quality of the generated DSM depends on the type and height of the forest and the spectral range used. In the visible wavelength range forest may be very dark; leading to very narrow histograms, while in the near infrared quite more contrast is available.

Table 7.2 Accuracy of height models based on space images in relation to reference DTMs. $rmse_{px}$ is the root mean square error of the x-parallax [GSD] (σ_z = B:H. σ_{px}) [Jacobsen, 2005 and 2006]

	GSD (m)	B:H	Area	σ_z (m)	σ_z as function of terrain inclination (m)	$rmse_{px}$ [GSD] flat area
TK350	(10/13)	0.5	Zonguldak open	23.3	20.0 + 23.9 . tan α	0.8
			forest	51.3	49.0 + 11.4 . tan α	1.9
			check points	6.6	4.7 + 2.2 . tan α	0.2
ASTER	15	0.6	Zonguldak open	25.0	21.7 + 14.5 . tan α	0.8
			forest	31.2	27.9 + 18.5 . tan α	1.1
			check points	12.7	–	0.5
KOMPSAT-1	6.6	0.5	Zonguldak open	13.6	11.3 + 11.5 . tan α	0.9
			forest	14.7	14.1 + 12.1 . tan α	1.1
SPOT-5	5	0.55	Zonguldak open	11.9	8.4 + 6.3 . tan α	0.8
			forest	15.0	9.8 + 5.3 . tan α	1.1
			check points	3.8	3.5 + 0.9 . tan α	0.4
SPOT-5 HRS	5 in flight direction 10 across	0.8	Bavaria open	6.7	6.4 + 4.9 . tan α	1.1
			forest	17.0	16.4 + 2.2 . tan α	2.7
SPOT-5 HRS filtered	5 in flight direction 10 across	0.8	Bavaria open	4.4	4.2 + 1.6 . tan α	0.7
			forest	12.3	10.0 + 6.9 · tan α	1.7
IKONOS	1	0.13	Maras open	1.7	same orbit	0.2
IKONOS	1	0.26	Zonguldak open	5.8	Δt = 3 month	1.5
QuickBird	0.62	0.11	Arizona open	4.8	Δt = 10 days	0.8
Cartosat-1	2.5	0.6	Mausanne open	4.0	3.9 + 1.6 . tan. α	1.0
			forest	3.6	3.3 + 0.3 . tan. α	0.8
Cartosat-1 filtered	2.5	0.6	Mausanne open	3.3	3.2 + 3.1 tan. α	0.8
			forest	3.5	2.9 + 1.8 tan. α	0.7

7.4 Global digital elevation models

7.4.1 GTOPO30

The only truly worldwide digital height model which is available free of charge is GTOPO30, generated by US Geological Survey's (USGS) EROS Data Center, completed in 1996 and later updated. Other "global" data sets do not give complete global coverage. GTOPO30 has a point spacing of 30 arcsec (approximately 926 m at the equator). It is based on several sources of topographic information, resulting in non-uniform accuracy. USGS estimates the accuracy with LE90 of 30 m, corresponding to a standard deviation of 18 m, but the accuracy varies greatly, depending on the location. The large spacing limits the use of this data set, allowing only a rough estimation of the Earth's surface.

7.4.2 Shuttle Radar Topography Mission

Digital elevation models have been generated for the land surface between latitudes 56° S and 62.5° N (Figure 7.20) by the Shuttle Radar Topography Mission (SRTM) [SRTM, 2011], using interferometric synthetic aperture radar (InSAR). The DSM, computed by means of the US C band SAR, is available with a spacing of 3 arcsec (approximately 92.6 m at the equator), free of charge via the Internet. It is also available with a spacing 1 arcsec, which is the spacing generated by the original process, but only for areas of the USA. DSMs from the German/Italian X band of the SRTM with a point spacing of 1 arcsec can be bought from the German Aerospace Center (DLR), but there are gaps between the ground tracks.

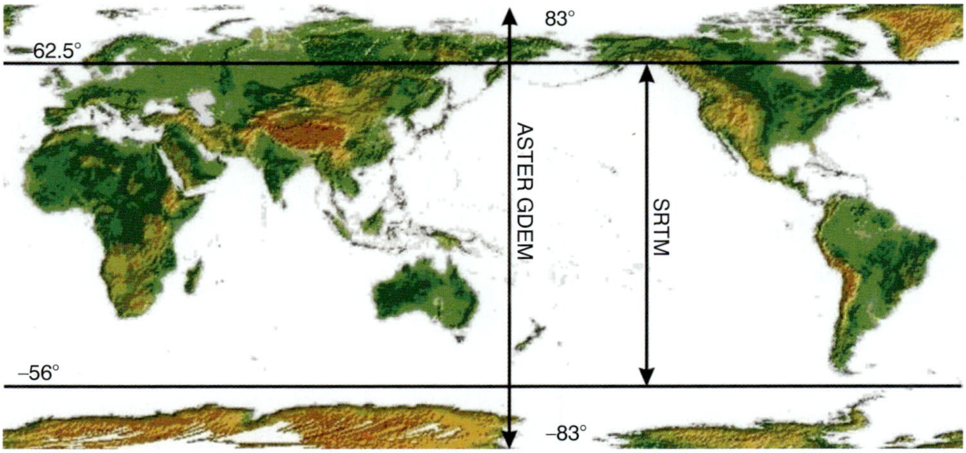

Figure 7.20 Area covered by SRTM C band DSM and ASTER GDEM.

The original SRTM DSM has some voids in steep mountainous areas, caused by viewing shadows and radar layover; in dry sand deserts, caused by lack of reflected energy; and on flat water surfaces, caused by specular reflection of the energy which is not reflected back to the sensor. Water surfaces can be filled by interpolation without problems. The voids in the SRTM C band DSM have been filled with a selection of non-radar data, but this leads to non-uniform accuracy.

7.4.3 ASTER GDEM

The ASTER GDEM, which began compilation in 2009, is another DSM that is free of charge. This covers the Earth between latitudes 83° S and 83° N (Figure 7.20), filling also gaps of the original SRTM DSM and having an improved spacing of 1 arcsec (approximately 31 m at the equator) [Abrams *et al.*, 2010]. ASTER GDEM is based on height models generated with images from the ASTER stereo satellite. All available ASTER height models for the same area have been merged. The ASTER GDEM webpage refers to the number of images used in generating the DSM as "stacks"; two stacks correspond to one height model.

As shown in Figures 7.21 and 7.22, the number of stacks used for ASTER GDEM varies considerably, depending on the location in the world [ASTER GDEM Validation Team, 2009], but also within the individual areas; the location influences the height accuracy as shown in Figures 7.21(b) and 7.22 [Jacobsen and Passini, 2010]. The adjusted lines represent the root mean square differences of ASTER GDEM heights with respect to reference DTMs. A general trend of increasing accuracy with a higher number of stacks can be seen; this is also influenced by vegetation and buildings. The SRTM DSM does not have such a variation in the individual tiles and has a more uniform character.

7.4.4 Assessment of ASTER GDEM and SRTM

ASTER GDEM and SRTM DSMs have been analysed by the ASTER GDEM Validation Team (2009), and more details have been added by Jacobsen and Passini (2010). For a comparison with reference DTMs, the horizontal fitting of the height models has to be checked and respected in advance. Besides datum problems in some countries, the comparison is influenced by an absolute scene orientation of ASTER and SRTM not based on ground control points. The root mean square coordinate error of the horizontal fit was determined to be 2.8 m for SRTM DSM and 7.2 m for the ASTER GDEM; the average bias in the height for 12 test areas was 3.5 m for SRTM DSMs and 6.2 m for ASTER GDEM. After a correct horizontal shift of the height models, a clear dependency on the terrain inclination can be seen, corresponding to $rmse_Z = A + B \cdot \tan \alpha$ (where α is terrain inclination). The factor B is 8.0 m on average for SRTM DSMs, and 6.7 m for ASTER GDEM.

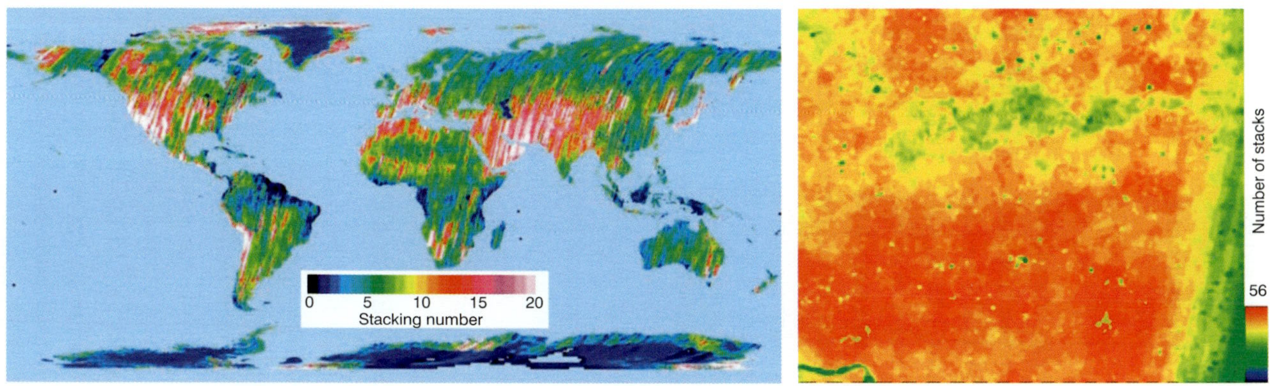

(a) (b)

Figure 7.21 (a) The number of stacks used for the ASTER GDEM over the globe [ASTER GDEM Validation Team, 2009]; and (b) the variation in the Pennsylvania test field (16 to 56 stacks/object points).

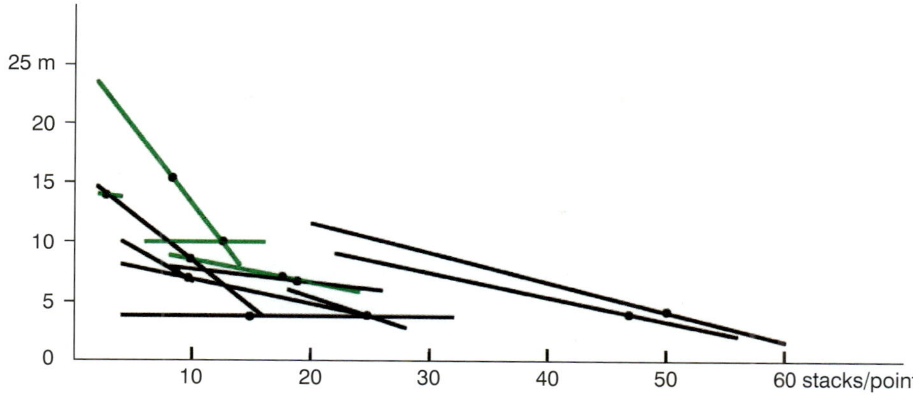

Figure 7.22 Root mean square differences of height in the ASTER GDEM, plotted against the number of stacks in 12 test areas for the range of available stacks/object point. Points on the lines are the average number of stacks for test areas used. Green lines show stacks influenced by forests.

Shifts of DSMs can be determined with a few control areas or a topographic map, making the standard deviation of the height, after correcting the bias, more realistic. A general problem of the quality estimation is the terrain coverage. The automatic generated height models based on optical images are related to the height of the visible surface, while SRTM C band InSAR shows the height of buildings and the height of vegetation slightly below the optical visible surface. By automatic filtering, at least some of the influence of the vegetation and buildings can be eliminated, leading to results closer to the bare ground.

The root mean square height differences with respect to reference DTMs of 12 worldwide distributed test areas with different characteristics, from flat to mountainous terrain, and with different coverage, from open to dense forest, are shown in Figure 7.23. The average root mean square error was determined to be

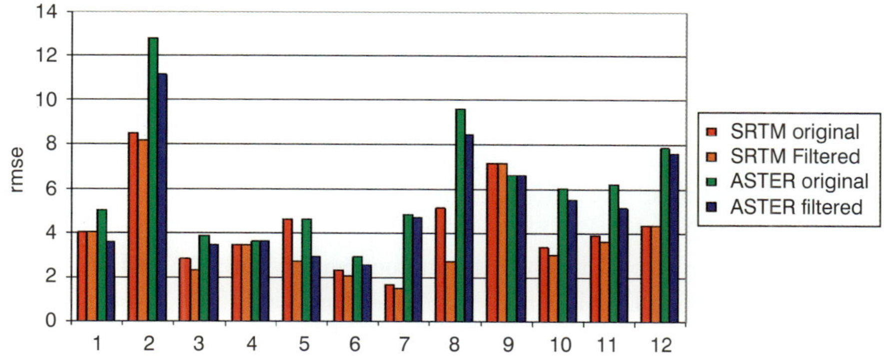

Figure 7.23 Root mean square errors of SRTM DSMs and ASTER GDEM in 12 test areas for flat terrain [Jacobsen and Passini, 2010].

5.6 m for SRTM DSMs and 8.2 m for ASTER GDEM; in flat terrain 5.1 m and 6.7 m, respectively, have been determined. Also in flat terrain, if the height models are filtered, the root mean square difference decreases to 4.3 m for SRTM DSMs and 5.8 m for ASTER GDEM. Test area 2 is almost completely covered by dense forest, and test area 9 has rough mountainous characteristics; if both these test areas are excluded, the root mean square height differences for flat terrain are reduced to 3.1 m for SRTM DSMs and 5.1 m for ASTER GDEM.

The geometric advantage of SRTM DSMs over ASTER GDEM is obvious, but ASTER GDEM has 1 arcsec point spacing in comparison to 3 arcsec for SRTM DSMs, so morphologic details should be clearer in ASTER GDEM. But morphologic detail in ASTER GDEM is actually no better than in SRTM DSMs. This may be caused by the fact that the averaged ASTER height models are not fitted to each other in *X* and *Y*, thus smoothing details. This has been improved with the second version, the ASTER GDEM2, available since October 2011.

7.4.5 Other DSMs

Astrium Geo-Information services produce a global DEM using stereo images from SPOT-5 HRS and TerraSAR data and is referred to as Elevation 3D. This is being developed by Spot Image in partnership with the French survey and mapping agency IGN. Reference3D is a georeferenced database containing three layers of information:

- DEM, with a spacing of 1 arcsec (~ 30 m at the equator), absolute elevation accuracy of 10 m and absolute planimetric accuracy of 15 m;
- GPS-compatible orthoimage;
- Quality and traceability metadata.

The Reference3D collection is ongoing, with large areas of the world already covered.

As an alternative to the free-of-charge height models, Intermap Technologies, Colorado, USA, have generated NEXTMap Europe and NEXTMap USA. These height models from airborne InSAR, covering 18 western European countries and the USA, have a point spacing of 0.15 arcsec, corresponding to approximately 5 m at the equator, and a standard deviation of the DSM of 1 m. Filtering has been done on NEXTMap data to produce a DTM.

The German Aerospace Centre (DLR) in cooperation with ASTRIUM GmbH will generate a worldwide DSM using TerraSAR-X and TanDEM-X, a constellation of two radar satellites. The imaging started in 2010 and will take two years to complete; in addition, it will require several years to process the data. TanDEM-X DSM will have a relative vertical standard deviation of 1.2 m, an absolute standard deviation of 6 m and a point spacing of 12 m.

7.4.6 Comparison of global DEMs

Table 7.3 Comparison of global DSMs

	ASTER GDEM	SRTM3 arcsec spacing	GTOPO30	TanDEM-X	Reference3D
Data source	ASTER	Shuttle Radar Topographic Mission	Variable	Interferometric SAR	SPOT-5 HRS
Interval	1 arcsec (~30 m)	3 arcsec (~90 m)	30 arcsec (~900 m)	12 m	1 arcsec (~30 m)
$rmse_z$	10 m	7 m	10–50 m	6 m	10 m
Relative $rmse_z$ (without bias)	8 m	6 m	10–50 m	1.2 m	15 m
DEM coverage	83° N to 83° S	60° N to 56° S	Global	Entire land surface of the Earth	83° N to 83° S
Area of missing data	Areas with no ASTER data due to constant cloud cover (supplied by other DSMs)	Topographically steep area, dry sand desert and water bodies (due to radar characteristics); version with void filling by other data available	None	None	Areas with no SPOT HRS coverage

7.5 Orthoimages and orthoimage mosaics

7.5.1 Orthoimage geometry

Orthoimages are one of the most often used photogrammetric products. An orthoimage is a projection of the image to a digital elevation model, leading to a product with the contents/radiometry of the image and the geometry of a map, without time-consuming manual mapping.

Geometric errors in an orthoimage are caused by errors in image orientation, the accuracy of the DTM and interpolation errors (Figure 7.24). The dominant error is usually caused by errors in the elevation model and by interpolation. Errors of the DTM cause position errors, depending on the tangent of the incidence angle *i*; in Figure 7.24, ΔP is a position error caused by an error in the DTM and is worse for larger incidence angles. Very high resolution space sensors have a small field of view but usually look in an off-nadir direction. Normal conditions for space images allow incidence angles of up to 18° to 30°: with an incidence angle of 30°, the position error corresponds to 57% of the height errors.

A smooth-looking orthoimage requires 8 pixels per mm (0.125 mm/pixel) in the projection scale. Corresponding to this specification, with 0.5 m GSD of WorldView or GeoEye-1, orthoimages with a scale number of 0.5 m/0.125 mm (4000), or a scale of up to 1:4000, can be generated. For orthoimages, as in the case of maps, 0.3 mm standard deviation of the position is acceptable, corresponding to 1.2 m or 2.4 GSD for images with a resolution of 50 cm GSD. The acceptable positional accuracy of 2.4 GSD is related to the above mentioned condition of 8 pixels per mm and can be applied independently of the GSD. For a standard deviation of the coordinate component of 1.2 m, a standard deviation of the height of 2.1 m is acceptable for images with 30° incidence angle, or 3.7 m for 18° incidence angle. As mentioned before, a standard deviation of the height in open areas, generated by automatic image matching, of 1 pixel times the height-to-base ratio, or 1.6 GSD under the condition of $B:H = 0.6$, is possible. This is better than that required for orthoimages, and does not cause problems if the height model is based on a satisfactory stereo model of images taken with the same sensor.

The real problems with orthoimages are caused by features not located on the height level of the DTM which is used, as is the case for buildings and bridges. The forest is not important in this relation. Initially, if a DSM based on space images is used, it fits to the top of the vegetation; in addition, it is not so important that a tree in the forest does not have a very precise location. The geometric problems of buildings can be seen in Figure 7.25(a), where the facades of a high building can be seen in a WorldView-1 image with 29.3° incidence angle. In maps, the building top can be seen and not the facades; furthermore, high buildings cause large occlusions.

With a true orthoimage, the correct geometry can be reached, but a detailed height model is required, including the height of any building corner in the upper and lower

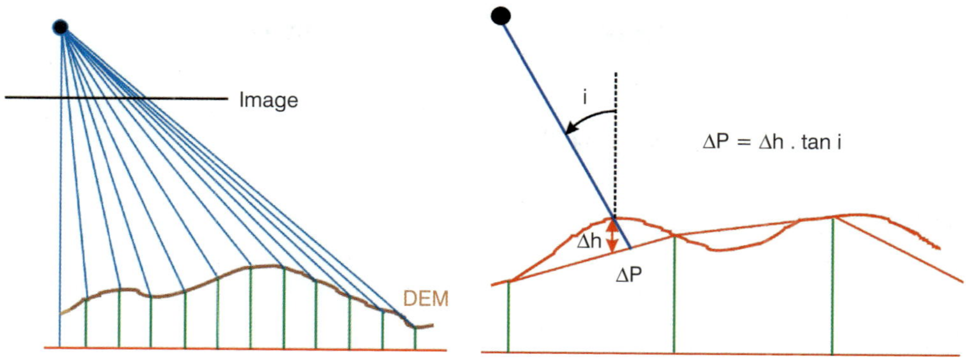

Figure 7.24 The principle of orthoimage geometry (left), and geometric errors due to DTM discrepancies and interpolation (right). ΔP is the position error caused by height discrepancies.

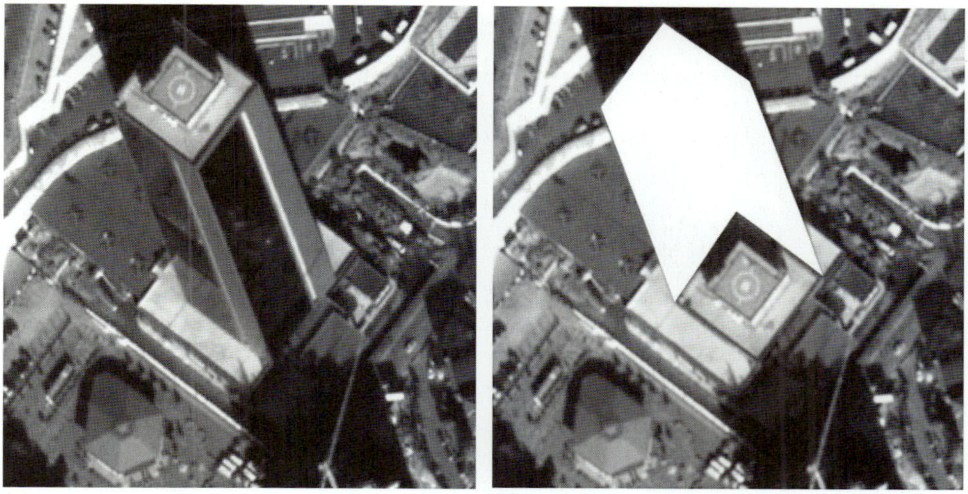

(a) (b)

Figure 7.25 (a) A building in a WorldView-1 image with 29.3° incidence angle; and (b) the correct location of building top – the white area is occluded in the original image. The dark area behind the building is shadow.

levels. Such a detailed height model can be generated by semi-global matching, but this method is more time consuming. Moreover, the occluded areas (Figure 7.25b) have to be inserted from other images, but in the case of space images, for financial and operational reasons, only one image (or just a few images) of the same area are available. Usually such displacements of buildings are acceptable, knowing that the geometry is correct only at the height of the DTM. For bridges this is different: if the DTM includes the height of the valley, the bridge is deformed, causing strange deformations as shown in Figures 7.26 and 7.27. Qualified orthoimages should be corrected manually for deformed bridges to avoid scepticism of orthoimages.

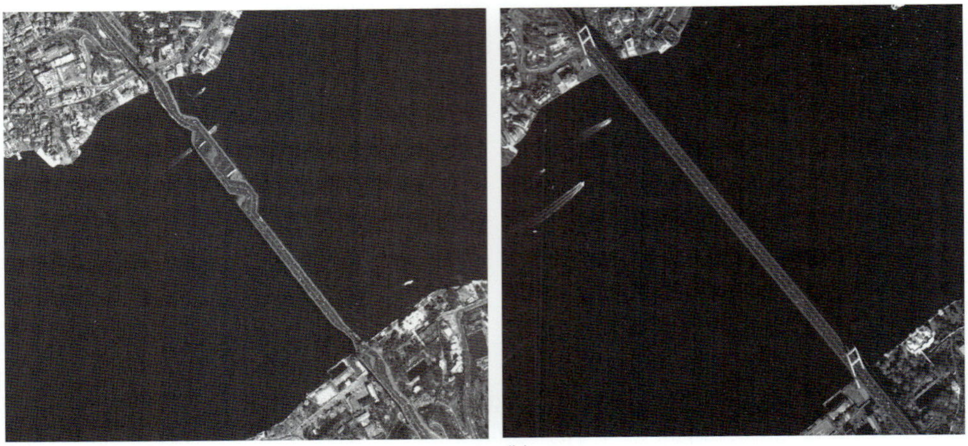

(a) (b)

Figure 7.26 (a) IKONOS orthoimage showing a deformed Bosporus bridge; and (b) original WorldView-1 image of the same bridge.

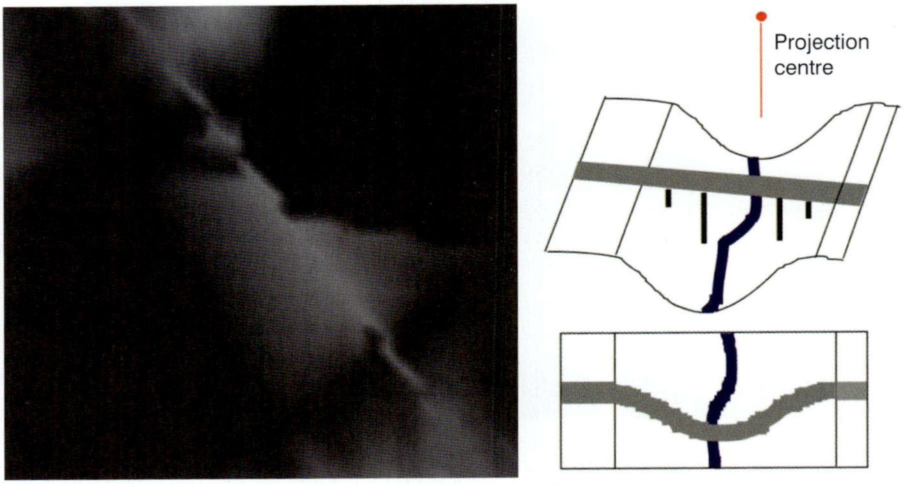

(a) (b)

Figure 7.27 (a) Grey value coded DTM used for the orthoimage shown in Figure 7.26; and (b) the principle of bridge deformation with DTM including height level of the valley.

7.5.2 Resampling

Orthoimages are generated with an indirect solution. Any output pixel has a defined location *X*, *Y*; the height is interpolated in the DTM at this position, leading to the *Z* value. Based on the image orientation, the image position for the ground coordinates *X*, *Y*, *Z* is computed. The grey value corresponding to the computed image position is used for the output pixel (Figure 7.28). The computed image position is usually not the centre of an input pixel, requiring the decision as to which grey value should be used.

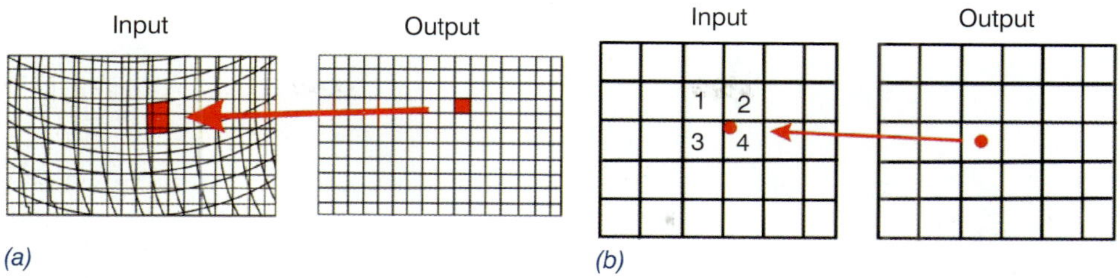

Figure 7.28 (a) Geometric principle of indirect orthoimage generation; and (b) problem of interpolation – the output grey level must be derived from four or more pixels of the input.

For this resampling, most commercial programs include the possibility of nearest neighbourhood, bilinear interpolation or cubic convolution.

In the case of nearest neighbourhood, the grey value of the nearest input pixel centre is used; in Figure 7.28 it is the grey value of pixel 4. Nearest neighbourhood resampling does not change the grey values, which is an advantage if the orthoimage is to be used for classification. In addition, it is quick, but has the disadvantage of deforming straight edges into staircases and reducing the accuracy by using the full pixel size instead of sub-pixel size.

In bilinear interpolation, the output grey value is linearly interpolated, depending on its precise position in relation to the neighbouring grey values. This includes a low computational cost and a relatively smooth output image. It improves the accuracy and the shape of edges. The interpretation of small details is better in orthoimages using bilinear interpolation than in images using the nearest neighbourhood (Figure 7.29 and 7.30), but the original grey values are changed and the image may appear slightly blurred.

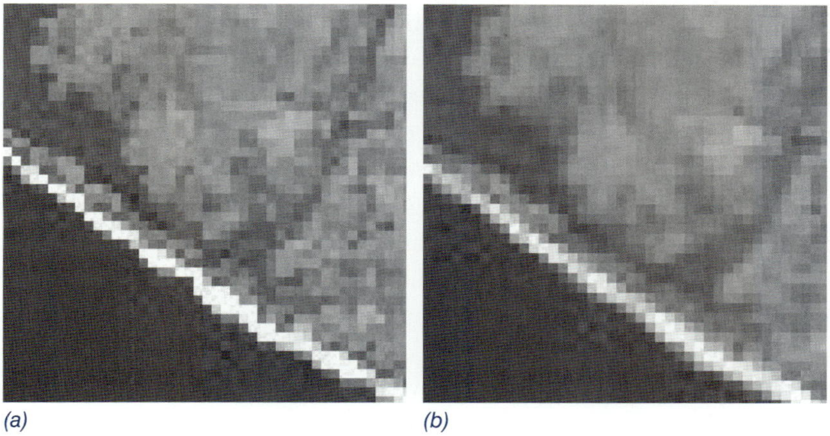

Figure 7.29 The appearance of a shoreline in a SPOT-5 orthoimage with: (a) nearest neighbourhood resampling; and (b) bilinear interpolation.

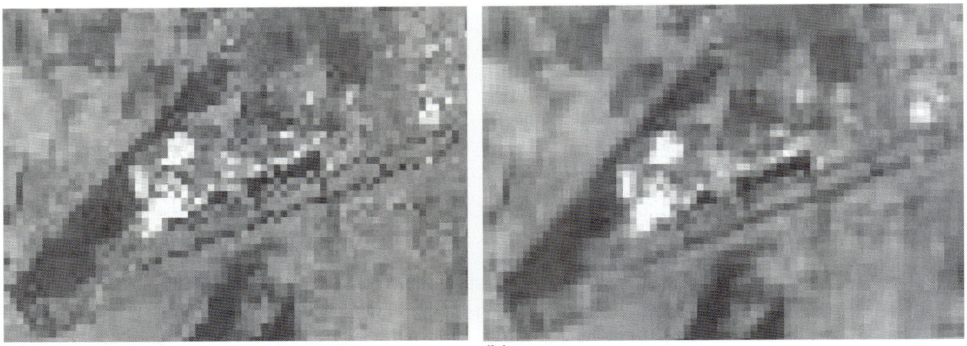

(a) (b)

Figure 7.30 The appearance of an airport in a SPOT-5 orthoimage with: (a) nearest neighbourhood resampling; and (b) bilinear interpolation.

The cubic convolution technique uses a matrix of neighbouring grey values. It is closer to the perfect $\sin(x)/x$ resampler than nearest neighbour or bilinear interpolation. The advantage of this method is an output image which is smooth and sharp; the disadvantage is a very high computational cost. The cubic convolution slightly enhances the contrast: this may be positive, but it is a change to the original appearance.

In Figure 7.29(a) the blocking effect of resampling by nearest neighbourhood is obvious; with bilinear interpolation the shoreline appears smoother. Figure 7.30 demonstrates the disadvantage of object interpretation by nearest neighbourhood, illustrating why resampling by nearest neighbourhood only should be used if the original grey values are important. Cubic convolution may be used if a slight contrast enhancement improves the image.

7.5.3 Image processing

Most optical satellite sensors have multispectral channels in addition to the panchromatic band. The fusion of lower resolution multispectral information with higher resolution panchromatic images is explained later. A standard user, not so much involved in remote sensing, prefers true colour images from red, green and blue channels (RGB), but not all satellite sensors include the blue spectral range. False colour infrared images (green, red, near infrared) have the advantage of showing more details of the vegetation, but with the disadvantage of looking different to what untrained users expect. These images can be converted to quasi-true colour images according to the transformation shown in Table 7.4.

The factor N has to be chosen for the simple image conversion. By default this should be equal to 1, but sometimes the images look more realistic with an enlarged factor, as in the case of the image of Figure 7.31(c), converted with $N = 20$ because the green colour dominates with $N = 1$. This conversion cannot generate real true colour images and also has some limitations, especially with the dominating green colour, but it is acceptable as an RGB image and simplifies the interpretation by the inexperienced user.

Table 7.4 The conversion of original input grey levels from multispectral channels to generate a quasi-true colour image

Output channel	Input grey value function
blue	green
green	(near-infrared grey value × 3 + N × green)/(3 + N)
red	red

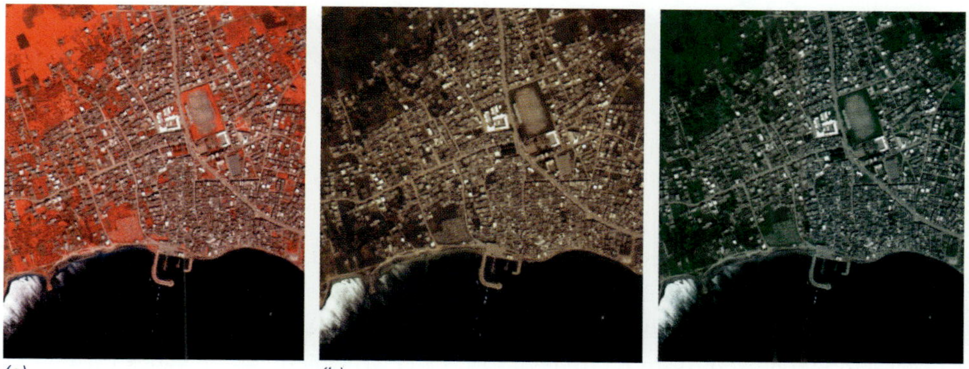

(a) (b) (c)

Figure 7.31 An IKONOS image: (a) original false colour infrared; (b) original RGB; and (c) false colour infrared computed RGB.

7.5.4 Generating stereo partners

Orthoimages do not contain displacement due to relief and so cannot be viewed stereoscopically. However, it is possible to generate a stereo partner, or stereo mate, which can be used together with the orthoimage to give a 3D view of the terrain. Orthoimages are formed by projecting the individual pixels from the image through the projection centre (which is different for any line) to the DTM, and from the DTM vertically to the coordinate reference datum. For the generation of a stereo partner, however, the projection from the DTM to the coordinate reference datum will not be done vertically, but inclined in the base direction (Figure 7.32).

The pixels in the generated stereo partner are linearly shifted with respect to the orthoimage, depending on the height of the DTM, corresponding to x-parallaxes in stereo models. A stereoscopic view of the combination of an orthoimage and a stereo partner will give a correct stereo impression and allows the determination of object heights. For the generation of a stereo partner, no stereo model is required if the height model for the generation comes from a different source. If the same image is used for the orthoimage and for the stereo partner, the stereo impression just shows the height

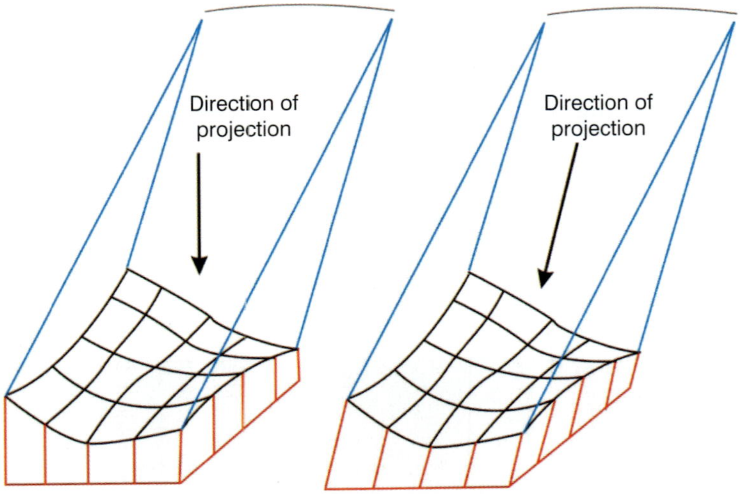

Figure 7.32 The geometry of an orthoimage (left) and stereo partner (right).

of the DTM which was used. If different images are used, details not included in the height level of the DTM, such as the height of buildings, can be seen stereoscopically.

Orthoimages are also the basis for computed 3D views. Orthoimages have the correct georeference and can be draped over a DTM, which can be viewed from any direction, leading to a computed stereo view, or it can be viewed as an image sequence for a fly-through.

7.5.5 Mosaicing orthoimages

Although images from space generally cover a large area, greater coverage is frequently required and for this it is necessary to mosaic orthoimages. It may also be necessary to fill gaps caused by clouds and haze. A typical requirement is shown as the red box in Figure 7.33(a). Different space images, even from the same sensor, have different radiometric properties caused by differences in sun elevation, view direction, haze, or colour of the vegetation, as seen in Figures 7.33(b) and (c). A simple overlay of neighbouring orthoimages on an orthoimage mosaic shows the boundary of the individual images clearly, disturbing the impression. The discrepancies should be limited to radiometric differences; geometric differences in open areas should not appear if the orthoimages have been generated correctly, but discrepancies cannot be avoided if there are height differences with respect to the height model used, as is the case with buildings (see below).

To give a pleasing appearance, generally grey value differences between neighbouring orthoimages can be modified by a linear grey value adjustment based on overlapping areas (Figure 7.33c); this can be successful if locally directed reflection is limited.

Seam lines along topographic linear features can also be used as shown in Figure 7.34(c). Seam lines along topographic lines can be combined with a linear grey value fit (Figure 7.34d). In most cases a radiometric normalisation is not desirable; only a local grey value fit between neighbouring images is used (Figure 7.35), with a satisfactory

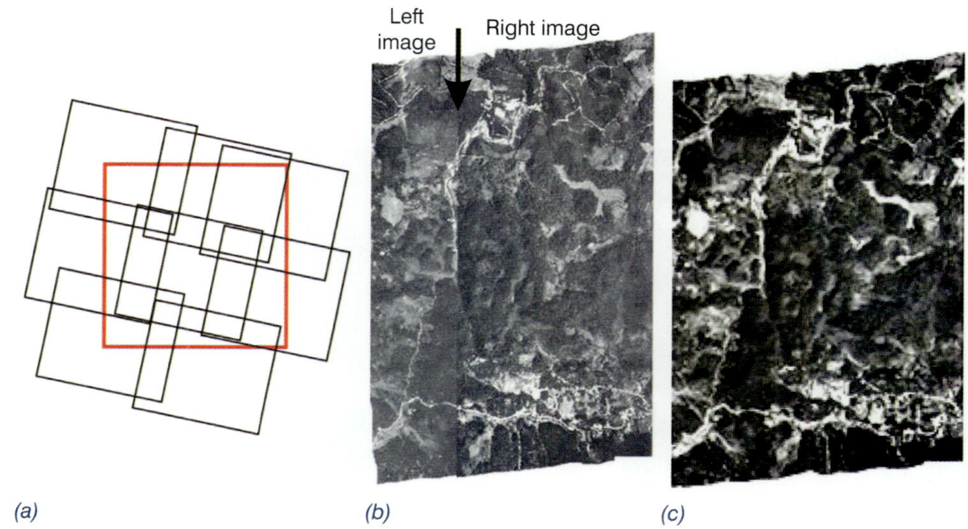

Figure 7.33 Mosaicing of orthoimages: (a) mosaic layout; (b) mosaic of two orthoimages based on images from the same day; and (c) after linear radiometric fit.

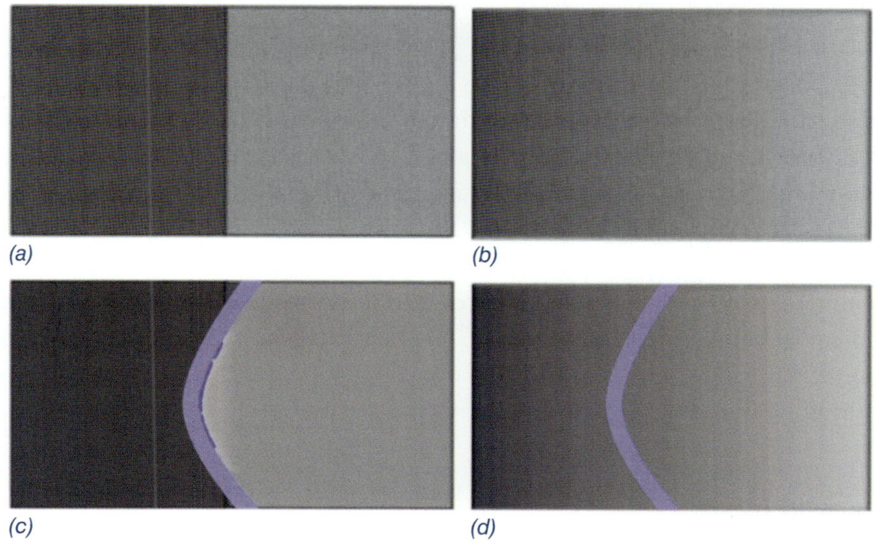

Figure 7.34 General principle of mosaicing: (a) original; (b) continuous grey level; (c) seam lines at linear topographic features; and (d) combination of both.

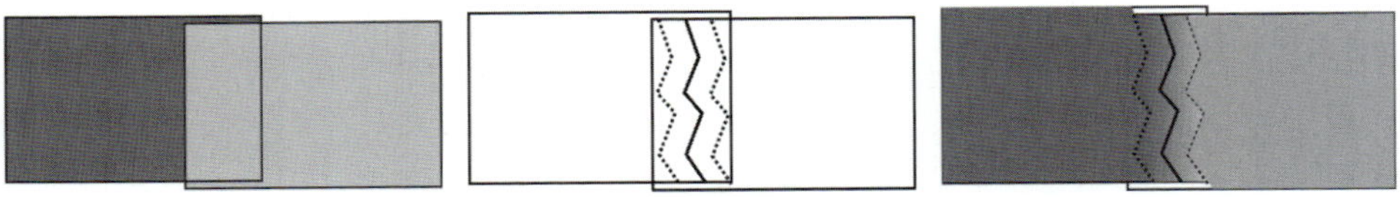

(a) (b) (c)

Figure 7.35 Handling seam lines with continuous grey value change: (a) overlapping orthoimages; (b) specified seam line with buffer zone; and (c) resulting grey value fit at seam line.

(a) (b)

Figure 7.36 A straight line cut of neighbouring aerial orthoimages at features not located at the level of the elevation model: (a) a building; and (b) a railway line.

buffer zone at the seam line [Afek and Brand, 1998]. The use of a transparent overlay of neighbouring orthoimages while changing the transparency from 0% up to 100% in the seam line area has the drawback of producing an unsharp impression, caused by a local misfit of neighbouring images.

A seam line along topographic features avoids the local mismatch of features, such as buildings and dams, not located at the level of the elevation model, highlighted in the example of aerial orthoimages in Figure 7.36. With very high resolution space images, starting at 1 m GSD, such problems also exist. Temporal changes, as in Figure 7.36(b) with a train only visible in one image, cannot be avoided.

7.6 Data fusion

7.6.1 Introduction

Data fusion allows different data sets to be merged to obtain higher quality information. A precondition for fusion is the synchronisation of the data: one element of one data set must have the correct geometric relation to the element of the other data set. Data fusion must follow the following steps:

1. Data alignment, which puts all data sets into the same reference system.
2. Data association, which puts the data into the same raster pattern.
3. Estimation, which determines the value to assign to each raster cell.

Applications related to optical satellites and their products are mainly the fusion of height models and the fusion of higher resolution panchromatic images with lower resolution colour images from the same optical system or even from different satellites.

7.6.2 Fusion of digital elevation models

Not only can digital elevation models with gaps be filled with other data in order to provide complete coverage, but also similar digital height models can be fused to reach higher accuracy and reliability. A typical example is the gap-filling of the original SRTM DSM (Section 7.4). SRTM DSMs have gaps in mountainous areas, caused by layover and viewing shadows (Figure 7.37a), in dry sand deserts (Figure 7.37b) and flat water surfaces. Before merging height models, the geometric relations have to be checked by comparing overlapping regions.

SRTM DSM gaps have been filled with different height models. The ASTER GDEM also has gaps, which have been filled mainly with SRTM DSM. This is also done for SPOT-5 HRS height models (Reference3D) to fill gaps, especially in dark forest areas. However, voids in one height model should not just be filled with height information from another source; first, the height models to be merged have to be geometrically registered. As mentioned in Section 7.4, SRTM DSMs have shifts relative

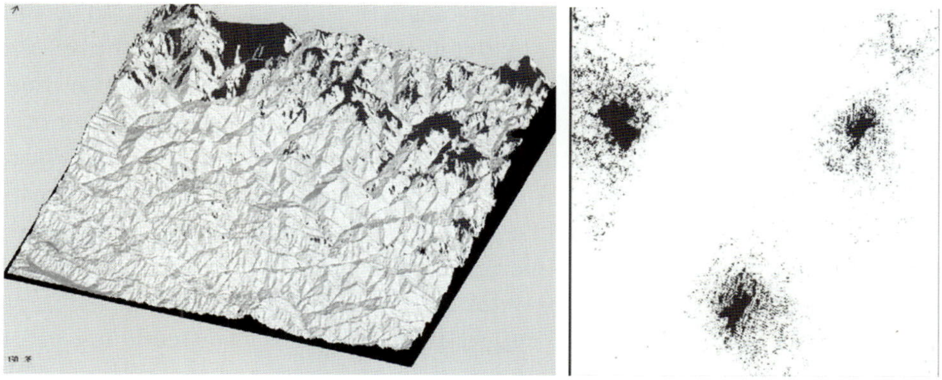

(a) (b)

Figure 7.37 Gaps in the original SRTM DSM 1° × 1° tiles are shown by the darkest shade: (a) mountainous area in the Himalayas; and (b) desert in Saudi Arabia.

to reference models in the region of 3.5 m, while the ASTER GDEM orientation errors range from 6 to 7 m in all three coordinate components. Such shifts cannot be neglected in relation to the system accuracy, so before merging the height models to fill gaps, shifts of the height models between each other have to be determined and corrected. The same problem of correct height model registration exists if similar height models are to be merged for accuracy improvement, such as in the case of ASTER GDEM. For ASTER GDEM, all available ASTER stereo models have been averaged. This simple averaging causes a loss of morphologic details, because the orientation errors of the individual ASTER height models have not been determined and taken into account. The next generation of ASTER GDEM will address this problem.

Honikel (1999) gives an example of how the fusion of a DSM generated from SPOT data can be fused with one from ERS SAR to improve the accuracy. Table 7.5 shows that fusion improves the overall result; fusion also significantly reduces the number of anomalous points, or spikes, in the DSM.

Table 7.5 Comparison of reference DTM with DSMs from SPOT, ERS and DSM SPOT fused with ERS

	SPOT	ERS	Fused
Mean difference (m)	1.9	1.0	1.4
rms difference (m)	7.9	4.7	4.0

7.6.3 Image fusion

Most optical Earth observation sensors provide images at high resolution in the panchromatic band and at lower resolution in the multispectral bands. For the small wavelength range of spectral bands, not as much energy comes through the optical system as through the large panchromatic range. In addition, the solar radiation is reduced in near infrared and especially in mid infrared, requiring a larger pixel size. Also the extremely high volume of image information needs to be reduced for transmission to the ground station. On the other hand, the human eye is not as sensitive to colour as it is to grey value information; a good compromise is to reconstruct colour images by pan-sharpening – the fusion of higher resolution panchromatic with lower resolution multispectral bands. Some sensors are limited to a panchromatic band that, for manual image interpretation, may be supported by colour information from another system. By fusion of the high resolution panchromatic image with lower resolution multispectral images from elsewhere, high resolution multispectral images can be generated.

As with the fusion of height models, the geometric relation of corresponding pixels in the different bands and images must fit. This is not a major problem if all the information originates from the same sensor, but orthoimages are necessary if images from different satellites are to be fused. Satellite systems such as IKONOS, QuickBird, WorldView-2 and GeoEye-1 have the CCD lines for panchromatic and multispectral bands in the same image plane (Figure 7.38). The offset between the panchromatic and multispectral CCD lines has to be respected, as well as the inner orientation, taking into consideration the geometric relation of the CCD lines to shifts, rotation and possible small curvature. Before fusion, the pixel size of the colour bands has to be changed to the pixel size of the panchromatic band. Nevertheless, sometimes a small misfit can be identified, indicated by colour bands in the pan-sharpened images, which should be compensated for by a shift of the corresponding scene.

The geometric co-registration is more complicated if images from different sensors or different orbits have to be merged. In such a case correct fusion is only possible with orthoimages. The geometric fit can be made with an approximation only for areas of flat or low relief and a small angle of convergence. By automatic image matching, tie points between the scenes can be determined and a transformation of one scene to the reference scene can be made separately for any tie-point triangle. A perspective or even an affine transformation may also be possible for flat areas.

Ehlers *et al.* (2008) classify image fusion by three levels: pixel level (iconic), feature level (symbolic) and knowledge or decision level. The pixel-level classification dominates the application and can be grouped into three classes: colour-related techniques,

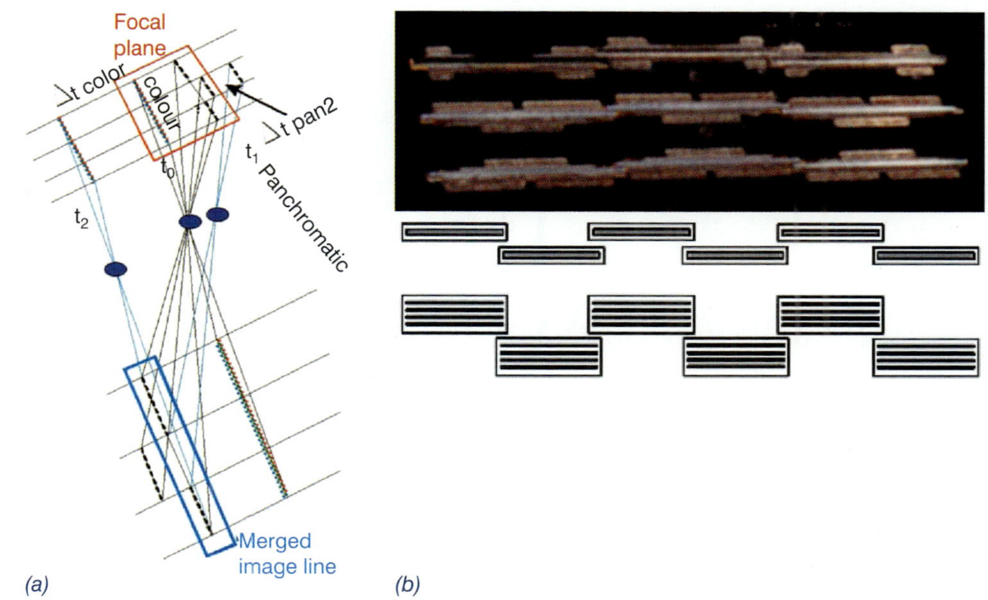

Figure 7.38 (a) Configuration of CCD lines in the IKONOS focal plane and their projection into the object space; and (b) CCD lines in IKONOS (top) – multispectral, panchromatic forward and backward scan, and the configuration of Quickbird CCD lines (bottom).

such as the IHS transformation; statistical methods, such as the principal component transform; and numerical methods, such as the Brovey transformation.

In the RGB colour model all possible colours are available by combining red, green and blue. The classical method of pan-sharpening is a transformation of three colour bands from the RGB colour model to intensity, hue and saturation (IHS); a replacement of the intensity channel by the panchromatic channel; and a reverse transformation to an RGB colour model (Figure 7.39). The steps in the process of pan-sharpening using IHS transformation are:

1. Geometric fit of the RGB image to the panchromatic image.
2. Multiplication of the number of multispectral pixels to fit the panchromatic pixels.
3. Transformation RGB → IHS.
4. Replacement of the intensity band by the panchromatic band.
5. Transformation IHS → RGB.

Using a principal component analysis (PCA), a transformation to a new colour space is possible, where one component includes the intensity, which can be replaced by the panchromatic band before reverse transformation. With PCA, several bands can be included in the pan-sharpening.

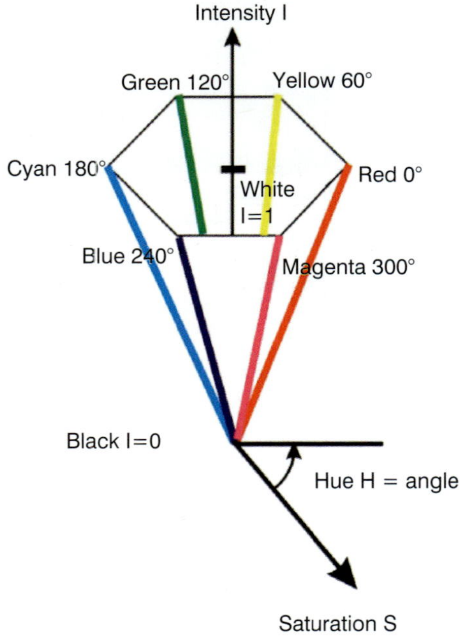

Figure 7.39 Relationship of RGB to IHS colour model.

$$RGB'_{ih} = \frac{RGB_{ih} \, . \, PAN_h}{RGB_{1h} + RGB_{2h} + RGB_{3h}} \tag{7.10}$$

Equation (7.10) is the Brovey fusion and no transformation to another colour model is required for pan-sharpening with this. The results are similar to the RGB-IHS-RGB transformation.

The standard pan-sharpening by IHS transformation, as well as by the Brovey fusion, creates spatially enhanced datasets, but usually at the expense of the spectral consistency [Ehlers *et al.*, 2008]. The spectral range of the panchromatic images often includes some parts of the near-infrared spectrum, leading to a change of the colour tone in pan-sharpened images. There are several methods of keeping the same colour balance as in the original RGB images [Cakir and Khorram, 2008]; nevertheless, the standard IHS transformation and the Brovey fusion give satisfactory results for a variety of uses.

Ehlers *et al.* (2008) developed a method for image fusion, preserving the spectral characteristics of the multispectral image, for both multi-date and multi-sensor data (Ehlers fusion), and compared it with other methods. Brovey, Ehlers, IHS, colour normalisation spectral sharpening and multiplicative fusion were found to produce good to very good spatial sharpening results. PCA and Gram Schmidt showed some improvement in spatial resolution but not to the extent demonstrated by the other methods. In some cases, the fused images still looked blurred. The quantitative analyses for colour constancy confirmed that Ehlers and wavelet fusion were always the best for all quality measurements employed.

7.7 Mapping

High resolution optical space imagery today competes with aerial images for topographic mapping. Images with GSD of 0.5 to 1 m may be acquired from space or from the air; whether to use space or aerial images is just a question of availability and economy. The coverage of an entire country with space images may be more expensive than with an aerial photography flight, but this may not be the case for smaller areas and countries without an infrastructure for aerial phototography flights. Furthermore, aerial images are classified in several countries, or permission to use existing aerial images or to undertake new photography flights is difficult or impossible to obtain.

The GSD is the dominant factor when selecting the type of imagery for mapping, as is obvious from Figure 7.40. With smaller GSD, more details can be identified. In addition, the colour plays a role: several mapping tests with space images showed that the object identification and classification is supported by the spectral information, but the final influence of the colour on the map contents is limited. Only very few additional

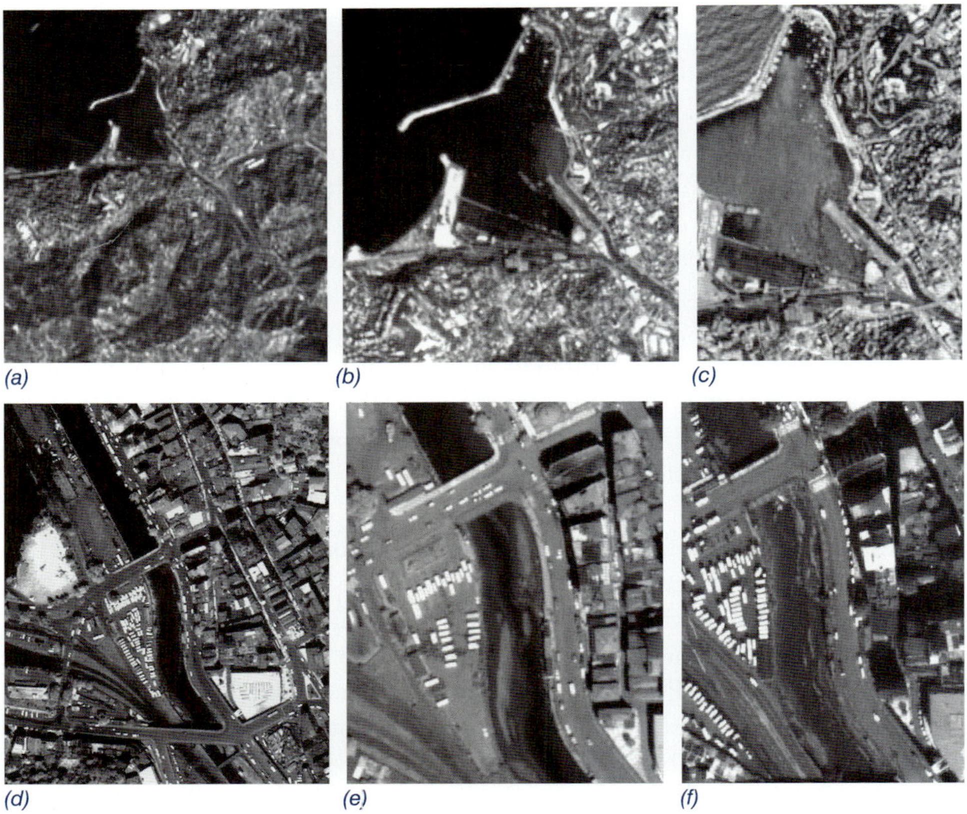

Figure 7.40 Comparison of the information content of different panchromatic space images of Zonguldak city: (a) ASTER, 15 m GSD; (b) IRS-1C, 5.7 m GSD; (c) SPOT-5, 5 m GSD; (d) IKONOS, 1 m GSD; (e) QuickBird, 0.6 m GSD; and (f) WorldView-1, 0.5 m GSD.

objects (less than 0.1%) can be identified by using colour instead of panchromatic images [Wegmann *et al.*, 1999; Topan *et al.*, 2004]. Also, trained operators will only make a very small number of misinterpretations in panchromatic images in relation to colour images, although colour images speed up object identification.

It is not only the ground resolution which is important: space images may have varying radiometric quality depending on the sensor, the atmospheric conditions and the sun elevation. It is also important that the GSD fits to the projected pixel size. For example, OrbView-3 has 1 m GSD, which is generated from staggered CCD lines, so the projected pixel size is 2 m. When determined by edge analysis and mapping tests, the images of such sensors as OrbView-3, IRS-1C/1D and SPOT-5 HRG supermode give effective GSDs corresponding to the information contents, of pixel sizes approximately 20% larger than the nominal GSD: 1.2 m, 6.8 m and 3 m for SPOT-5 HRG, respectively.

Sometimes the atmospheric conditions are not optimal because of haze and dust. Clouds and haze affect object interpretation, and it is often difficult to identify objects in cloud shadows. The sun elevation plays an important role as well. As shown in Figure 7.6, shadow deteriorates information. The influence of different shadow lengths and variable atmospheric conditions for mapping the city of Zonguldak is evident in Figure 7.41. In the case of 41° sun elevation the streets were hidden and the atmosphere had very little haze, leading to an insignificant amount of stray light, which caused very dark shadows. If a Wallis filter were to be used here, no details of the streets would be seen and it would not be possible to map walkways; this is not the same as the situation shown in Figure 7.6. Information in shadows can only be improved using a local Wallis filter if there is some information there in the first place. If the atmosphere is very clear, no stray light illuminates the areas of shadow, and objects cannot be identified. In the example shown in Figure 7.41, the classification of streets and backyards may even be confused.

The relationship between the scale of topographic mapping that is possible and GSD depends on the mapping accuracy and the information content. A standard deviation of horizontal coordinate components of 0.3 mm at the map scale is acceptable for topographic maps. Mapping of well-specified objects is possible with a standard deviation of the coordinates of 1 GSD, leading to a mapping accuracy requirement of 0.3 mm, corresponding to 1 GSD, which equates to a map scale number of 1 GSD / 0.3 mm. For example, for IKONOS with 1 m GSD, a map scale number of 1 m / 0.3 mm = 3333 would be possible to yield acceptable accuracy. But in fact the limiting factor is the information content. By comparing the information content included in space as well as in aerial images of different GSD in relation to topographic maps, the empirical requirement of 0.05 mm to 0.1 mm GSD in the map has been determined [Jacobsen and Büyüksalih, 2008]. In the case of IKONOS images with 1 m GSD, this corresponds to a possible topographic map scale of 1:20 000 to 1:10 000. GeoEye published a table showing map scales just related to the accuracy attainable and not respecting the more important requirements for object identification [GeoEye, 2011], leading to results which are too optimistic. In addition, some important details for mapping cannot be identified with a GSD of less than 5 m.

(a) (b) (c)

Figure 7.41 IKONOS images of Zonguldak, taken at different sun elevations: (a) 46°; (b) 41°; and (c) 61°.

The relationship between GSD and possible scale of topographic maps is shown in Figure 7.42. In reality, the boundaries are not so distinct; the relationship also depends on the type of area. Objects may be close together, arranged irregularly and partly hidden by trees, as shown in the IKONOS image of Zonguldak (Figure 7.43a); they may be well defined, as shown in the IKONOS image of Dubai (Figure 7.43b); or they may be large with space in between, as shown in the QuickBird image of Atlanta (Figure 7.43c).

In addition, the content of topographic maps is not fixed: specifications vary between countries. In the United States, only limited details are included in

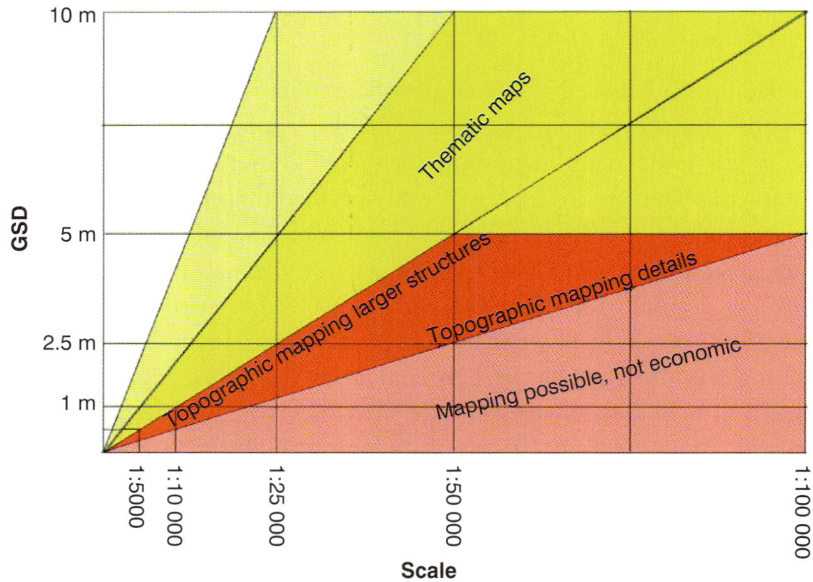

Figure 7.42 Relationship between the scale of a topographic map and GSD [Jacobsen and Büyüksalih, 2008].

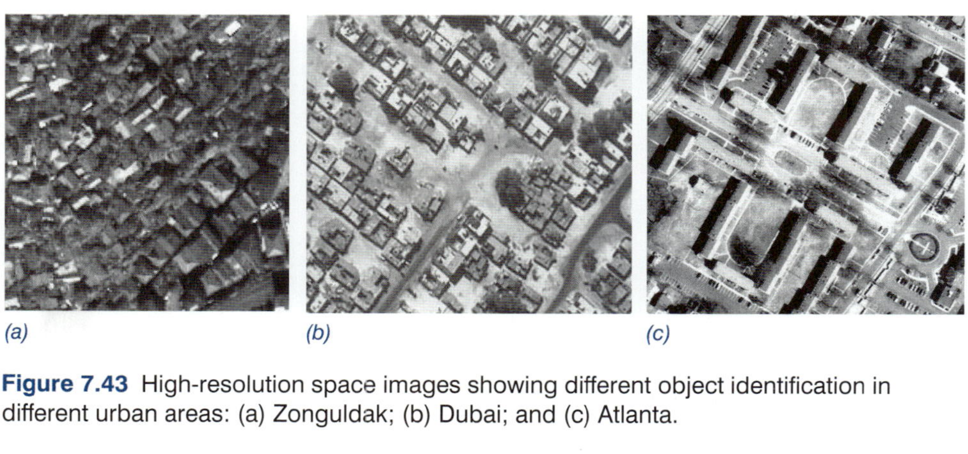

(a) *(b)* *(c)*

Figure 7.43 High-resolution space images showing different object identification in different urban areas: (a) Zonguldak; (b) Dubai; and (c) Atlanta.

topographic maps, while a lot of details are shown in Switzerland. This leads to a variation of the relationship between GSD and possible map scale as shown in Figure 7.42. For thematic maps the larger range depends on the topic and here no clear limit can be proposed.

NATO specifies the required GSD for different object types, according to three classes: detection, recognition and identification (Table 7.6). For elements included in topographic maps, at least "recognition" is required; in urban areas, individual houses correspond to "identification".

Another specification for topographic maps has been given by Wegmann *et al.* (1999) and shown in Table 7.7. Thus, for example, extensions to standard size buildings are shown in maps of 1:5000 scale and larger, while individual buildings are included in topographic maps of 1:20 000 scale and larger, confirming the relationship in Figure 7.42, including the limit of 5 m GSD for topographic mapping. Topographic maps have been generated from SPOT 1–4 images with 10 m GSD, but these maps have never been sufficiently complete to satisfy requirements for topographic maps compiled to normal standards, and have had to be supported by intensive ground surveys.

Table 7.6 Required ground pixel size for optical image interpretation, corresponding to NATO standards (STANAG 3769)

Object	Detection (m)	Recognition (m)	Identification (m)
Terrain features	~800	90	3
Urban areas	60	15	3
Roads	6	4.5	1.5
Railways	15	4.5	1.5
Bridges	6	4.5	1.5
Airfield facilities	6	4.5	3

Table 7.7 Required GSD for topographic mapping [Wegmann *et al.*, 1999]

	Required GSD (m)
Urban buildings	2
Building extensions	0.6–1
Foot path	2
Minor road network	5
Railway	5
Fine hydrology	5
Major road network	10
Blocks of buildings	10

For instance, a mapping test in Nigeria with 10 m GSD did not include the major roads, because they were partially hidden by trees.

Tests of mapping built-up areas have shown that it is difficult to map building extensions and the correct rotation of buildings, as presented in Figure 7.44(a) with original QuickBird colour images having 2.4 m GSD. In this case the resolution of the panchromatic band with 0.61 m GSD is required (Figure 7.44b). With 1 m GSD images of the same area taken by IKONOS, the mapping of major building extensions

(a) (b) (c)

Figure 7.44 Mapping of buildings with QuickBird images: (a) colour images with 2.44 m GSD; (b) panchromatic images with 0.61 m GSD; and (c) superimposed vectors from 2.44 m GSD (green lines) and 0.61 m GSD (red lines).

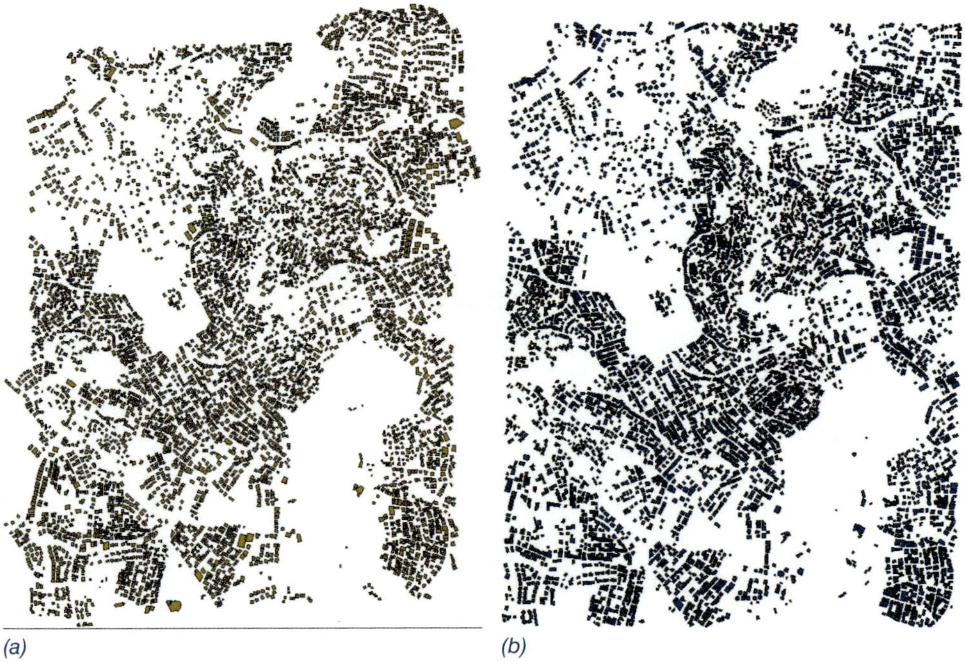

(a) (b)

Figure 7.45 (a) Mapping buildings in Istanbul with a WorldView-1 image; and (b) reference map from an aerial survey.

is possible. BIMTAS, the mapping company of the Greater Istanbul Municipality, used IKONOS pan-sharpened images for the update of buildings in maps [Büyüksalih and Jacobsen, 2008]. New buildings identified in the images have been checked by ground inspections.

BIMTAS also successfully used WorldView-1 images of 0.5 m GSD to map buildings in Istanbul [Büyüksalih *et al.*, 2009] as shown in Figure 7.45. It can be seen that the same information as from aerial photographs could be acquired from these images. All buildings were detected with the advantage of up-to-date information. Tests with SPOT-5 panchromatic scenes (5 m GSD), as well as IRS-1C (5.7 m GSD) and KOMPSAT-1 (6.6 m GSD), confirmed the findings: with approximately 5 m GSD the road network in densely built-up areas can be mapped with few exceptions, as well as in urban areas without details. If the effective GSD, taking account of the image quality determined by edge analysis, exceeds 5 m GSD, some important roads cannot be identified.

Automatic object extraction requires more detailed images than mapping by experienced human operators. Operational applications are still limited today, even if based on aerial information. Some impressive results have been published, but for real operational applications an intensive manual check and support is required. One exception for operational use is the development of the system WiPKA by the Leibniz University Hannover [Gerke and Heipke, 2008], used by the BKG, Frankfurt, Germany, for the verification of road information. WiPKA uses a traffic-light system in which reliable extracted roads are shown in green, while unreliable lines are shown in red, for manual inspection. The system is also successful when using ALOS PRISM images [Fujimura *et al.*, 2008]. Other information, for example from agricultural areas, can be extracted [Helmholz and Rottensteiner, 2009], but the reliability is limited.

References

Abrams, M., Bailey, B., Tsu, H. and Masami, H., 2010. The ASTER global DEM. *Photogrammetric Engineering and Remote Sensing*, 76(4), 344–348.

Afek, Y. and Brand, A., 1998. Mosaicking of orthorectified aerial images. *Photogrammetric Engineering and Remote Sensing*, 64(2), 115–125.

Alobeid, A., Jacobsen, K. and Heipke, C., 2009. Building height estimation in urban areas from very high resolution satellite stereo images. *International Archives of the Photogrammetry, Remote Sensing and Spatial Information Sciences*, 38(1-4-7/W5), http://www.isprs.org/proceedings/XXXVIII/1_4_7-W5

Alobeid, A., Jacobsen, K., Heipke, C., Alrajhi, M., 2011. Building Monitoring with Differential DSMs. IntArchPhRS. *International Archives of the Photogrammetry, Remote Sensing and Spatial Information Sciences*, 38 (4/W19).

ASTER GDEM Validation Team, 2009. METI/ERSDAC, NASA/LPDAAC, USGS/EROS, 2009: ASTER Global DEM Validation, Summary Report, http://www.ersdac.or.jp/GDEM/E/image/ASTERGDEM_ValidationSummaryReport_Ver1.pdf (accessed March 2011).

Baltsavias, E., Pateraki, M. and Zhang, L., 2001. Radiometric and geometric evaluation of IKONOS geo images and their use for 3D building modelling. Joint ISPRS Workshop High Resolution Mapping from Space, Hannover, Germany, http://www.photogrammetry.ethz.ch/general/persons/maria/pub/hannover.pdf (accessed March 2011).

Baudoin, A., Schroeder, M., Valorge, C., Bernard, M. and Rudowski, V., 2004. The HRS-SAP initiative: A scientific assessment of the high resolution stereoscopic instrument on board of SPOT 5 by ISPRS investigators. *International Archives of the Photogrammetry, Remote Sensing and Spatial Information Sciences*, 35(B1), 372–378.

Birchfield, S. and Tomasi, C., 1998. A pixel dissimilarity measure that is insensitive to image sampling. *IEEE Transactions on Pattern Analysis and Machine Intelligence*, 20(4), 401–406.

Birchfield, S. and Tomasi, C., 1999. Depth discontinuities by pixel-to-pixel stereo. *International Journal of Computer Vision*, 35(3), 269–293.

Büyüksalih, G. and Jacobsen, K., 2008. Determination of the build up area development in the greater municipality of Istanbul by space images. EARSel Workshop Remote Sensing – New Challenges of High Resolution, Bochum, Germany, http://www.ipi.uni-hannover.de/uploads/tx_tkpublikationen/BuildUp_Bochum.pdf (accessed February 2011).

Büyüksalih, G., Baz, I., Bayburt, S., Jacobsen, K. and Alkan, M., 2009. Geometric mapping potential of WorldView-1 images. *International Archives of the Photogrammetry, Remote Sensing and Spatial Information Sciences*, 38(1-4-7/W5), http://www.isprs.org/proceedings/XXXVIII/1_4_7-W5/paper/papers.htm (accessed March 2011).

Cakir, H.J. and Khorram, S., 2008. Pixel level fusion of panchromatic and multispectral images based on correspondence analysis. *Photogrammetric Engineering and Remote Sensing*, 74(2),183–192.

Ehlers, M., Klonus, S. and Astrand, P.J., 2008. Quality assessment for multi-sensor multi-date image fusion. *International Archives of the Photogrammetry, Remote Sensing and Spatial Information Sciences*, 37(B4), 499–505.

Fujimura, H., Ziems, M. and Heipke, C., 2008. De-generalization of Japanese road data using satellite imagery. *Photogrammetrie, Fernerkundung und Geoinformation*, 2008(5), 363–373.

GeoEye, 2011. http://www.geoeye.com/CorpSite/products-and-services/imagery-collection/satellite-imagery-products/product-specifications.aspx (accessed September 2011).

Gerke, M. and Heipke, C., 2008. Image based quality assessment of road databases. *International Journal of Geoinformation Science*, 22(8), 871–894.

GTOPO30. http://www1.gsi.go.jp/geowww/globalmap-gsi/gtopo30/gtopo30.html (accessed August 2010).

Heipke, C., Piechullek, C. and Ebner, H., 2001. Simulation studies and practical tests using multi-image shape from shading. *ISPRS Journal of Photogrammetry and Remote Sensing*, 56, 139–148.

Helmholz, P. and Rottensteiner, F., 2009. Automatic verification of agriculture areas using IKONOS satellite images. *International Archives of the Photogrammetry, Remote Sensing and Spatial Information Sciences*, 38(1-4-7/W5), http://www.isprs.org/proceedings/XXXVIII/1_4_7-W5/paper/papers.htm (accessed March 2011).

Hirschmüller, H., 2008. Stereo processing by semiglobal matching and mutual information. *IEEE Transactions on Pattern Analysis and Machine Intelligence*, 30(2), 328–341.

Hirschmüller, H. and Scharstein, D., 2009. Evaluation of cost functions for stereo matching. *IEEE Conference on Computer Vision and Pattern Recognition*, 31(9), 1582–1599.

Honikel, M., 1999. Strategies and methods for the fusion of digital elevation models from optical and SAR data. *International Archives of Photogrammetry and Remote Sensing*, 32(7-4-3 W6).

Jacobsen, K., 2002. Mapping with IKONOS images. In Benes, T. (ed.), *Geoinformation for European-wide Integration*, Millpress, Rotterdam, Netherlands, 149–156. 22nd EARSeL Symposium on Remote Sensing, Prague, Czech Republic, http://www.ipi.uni-hannover.de/uploads/tx_tkpublikationen/jacmwi.pdf (accessed March 2011).

Jacobsen, K., 2003. Analysis of SPOT HRS stereo data. Joint Workshop High Resolution Mapping from Space 2003, Hannover, Germany, http://www.ipi.uni-hannover.de/uploads/tx_tkpublikationen/jacobs.pdf (accessed March 2011).

Jacobsen, K., 2005a. DEMs based on space images versus SRTM height models. ASPRS Annual Convention, Baltimore, MD, USA, http://www.ipi.uni-hannover.de/uploads/tx_tkpublikationen/jac_SRTM_SI2005.pdf (accessed March 2011).

Jacobsen, K., 2005b. Analysis of SRTM elevation models. EARSeL 3D-Remote Sensing Workshop, Porto, Portugal, http://www.ipi.uni-hannover.de/uploads/tx_tkpublikationen/ASEjac.pdf (accessed March 2011).

Jacobsen K., 2006. ISPRS-ISRO Cartosat-1 scientific assessment programme (C-SAP) technical report – test areas Mausanne and Warsaw. *International Archives of the Photogrammetry, Remote Sensing and Spatial Information Sciences*, 36(4), 1052–1056.

Jacobsen, K., 2008. High resolution optical satellite images – from scene orientation to GIS-data acquisition. INCA congress, Gandhinagar, India, http://www.ipi.uni-hannover.de/uploads/tx_tkpublikationen/INCA_Jacobsen08.pdf (accessed February 2011).

Jacobsen, K. and Büyüksalih, G., 2008. Mapping from space for developing countries. EARSel Workshop Remote Sensing – New Challenges of High Resolution, Bochum, Germany, http://www.ipi.uni-hannover.de/uploads/tx_tkpublikationen/Mapping_Bochum.pdf (accessed March 2011).

Jacobsen, K. and Passini. R., 2010. Analysis of ASTER GDEM elevation models. *International Archives of the Photogrammetry, Remote Sensing and Spatial Information Sciences*, 38(1), http://www.isprs.org/proceedings/XXXVIII/part1/09/09_03_Paper_103.pdf (accessed March 2011).

Passini, R., Betzner, D. and Jacobsen, K., 2002. Filtering of digital elevation models. ASPRS Annual Convention, Washington, D.C., USA, http://www.ipi.uni-hannover.de/uploads/tx_tkpublikationen/fltpbj02.pdf (accessed March 2011).

Passini, R. and Jacobsen, K., 2007a. Accuracy analysis of SRTM height models. ASPRS Annual Conference, Tampa, FL, USA, http://www.ipi.uni-hannover.de/uploads/tx_tkpublikationen/Passini_jac.pdf (accessed March 2011).

Passini, R. and Jacobsen, K., 2007b. High resolution SRTM height models. *International Archives of the Photogrammetry, Remote Sensing and Spatial Information Sciences*, 36(I-W51), http://www.isprs.org/proceedings/XXXVI/1-W51/paper/papers.htm (accessed September 2011).

Pohl, C. and van Genderen, J.L., 1998. Multisensor image fusion in remote sensing: Concepts, methods and applications. *International Journal of Remote Sensing*, 19(5), 823–854, http://www.scribd.com/doc/38517632/Multisensor-image-fusion-in-remote-sensing-concepts-methods-and-applications (accessed March 2011).

SRTM, 2011. http://www2.jpl.nasa.gov/srtm (accessed February 2011).

Topan, H., Büyüksalih, G. and Jacobsen, K., 2004. Comparison of information contents of high resolution space images. *International Archives of the Photogrammetry, Remote Sensing and Spatial Information Sciences*, 35(B4), 583–588.

Toutin, T., 2005. Generation of DTM from stereo high resolution sensors. *International Archives of the Photogrammetry, Remote Sensing and Spatial Information Sciences*, 36(1/W3). ISPRS Workshop – High-Resolution Earth Imaging for Geospatial Information, Hannover, Germany, http://www.ipi.uni-hannover.de/fileadmin/institut/pdf/032-toutin.pdf (accessed March 2011).

Wegmann, H., Beutner, S. and Jacobsen, K., 1999. Topographic information system by satellite and digital airborne images. Joint Workshop ISPRS Working Groups I/1, I/3 and IV/4 – Sensors and Mapping from Space, Hannover, Germany, http://www.ipi.uni-hannover.de/uploads/tx_tkpublikationen/wegmann.pdf (accessed February 2011).

Zhang. L. and Gruen, A., 2006. Multi-image matching for DSM generation from IKONOS imagery. *ISPRS Journal of Photogrammetry and Remote Sensing*, 60, 195–211.

Conclusions and Future Developments | 8

Ian Dowman

8.1 Summary

8.1.1 Status of sensors

There are two aspects to the current status of high and very high resolution sensors. First, there is a stable situation with large commercial systems, whereby resolution is slowly increasing and the speed at which images can be acquired is also increasing; second, there is a rapidly developing small-satellite market, with sensors moving towards a similar resolution to that of larger systems, but with limited imaging capacity. The market for high resolution imagery is expanding, but current systems, from four major companies, are able to satisfy the demand. The use of higher resolution for commercial systems is limited by restrictions imposed by the government of the United States. Today Earth observation systems, including the ground stations, can be ordered almost off the shelf. The price of components is lower than in the past, and the components are becoming more compact. All this reduces the total cost.

Nevertheless, large mirror telescopes are required for a ground resolution of 1 m or less. For a high imaging capacity, a sufficient on-board storing capacity is required, as well as considerable satellite agility. For better agility there is a tendency to move away from reaction wheels to the more powerful control moment gyros. Downlink capacity is still the bottleneck, requiring highly developed image compression techniques. These limitations do not seem to affect the demand for satellites for national and regional use, and there are examples of systems designed and operating to serve national requirements, for example CBERS and RazakSAT. Commercial operators are also using small satellites, as in the case of the RapidEye constellation. Resolution is increasing, with Surrey Satellite Technology Limited (SSTL) having designed a small satellite with 1 m GSD. The use of constellations, for example RapidEye, is also enabling shorter revisit times to almost any point on the planet.

There are more exciting developments in the area of radar, with new systems such as RadarSat, TerraSAR-X, TanDEM-X and COSMO-SkyMed already launched and small radar sensors being developed by, for example, SSTL.

8.1.2 Processing and applications

Software for processing is now well developed and operational within most digital photogrammetric software packages, requiring minimum understanding by operators. Orthoimages and digital elevation models can be produced very easily because of good matching software and the ability to overlay an image onto the DEM. The major companies offer value-added products such as orthoimages, and companies like RapidEye offer services such as agricultural monitoring and change detection. The main application areas are:

- monitoring for many objectives, such as urban change, mapping, agriculture, forestry and disaster management;
- surveillance for national security and illegal activity, for example at sea;
- provision of data for global data sets, such as Google Earth and Bing, and for updating databases used for satellite navigation and location-based services;
- provision of images to the media.

Earth observation data has proved its importance in disaster management, and the way in which it can be used is often demonstrated, for example in response to the Haiti Earthquake which occurred at 16.53 local time on 12 January 2010. An alert was given 27 minutes after the earthquake, and response activation of UN-SPIDER, UN-Spaceaid and the International Charter Space and Major Disasters occurred 4 hours 35 minutes after the event. The first satellite image covering Port-au-Prince was acquired by GeoEye on 13th January and delivered by Google Crisis Response Team as KML and a base layer in Google Map/Earth. DigitalGlobe (WorldView 1–2, QuickBird) covered the whole country between 13th and 17th January. Data was delivered as WMS, KML and GeoTiff through the DigitalGlobe Crisis Event Services. High resolution radar imagery was acquired by COSMO-SkyMed and TerraSAR-X. Aerial surveys using panchromatic and thermal images and lidar started on 21st January and covered most of the affected areas. By 31st January, 324 maps had been produced, 43 of which were posted on ReliefWeb. This demonstrates the versatility of high resolution satellite data (including radar), how satellite operators can work together to acquire data very quickly and the extent to which delivery is also important in disaster situations.

8.1.3 Organisational and administrative issues

There are limits on the development of hardware as summarised above, but there are also limits imposed by government regulation. Companies in the United States may not distribute products with a GSD of less than 0.5 m, and although sensors are collecting data at higher resolutions, these are only available to government agencies. Other countries, such as India, restrict distribution of high resolution data of their country. If resolution were to improve further, additional restrictions could be applied. The government also influences the cost of data, in that the US companies are contracted to provide data, often as a priority, for defence purposes; this acts as a subsidy, which enables the development of better sensors. As an example, the cost of absolute positioning to better than 15 m is very high, but this is possible with the US commercial sensors as well as with RapidEye, and is proposed for Pleiades, SPOT-6 and SPOT-7. The market also affects the viability of sensors: customers may be prepared to pay for images, but are reluctant to pay for costly value-added products. There is already competition between satellite imagery and aerial imagery at 0.5 to 1 m GSD; if higher resolution satellite data were to become available, the competition would be greater.

There is also political influence on development. The Global Monitoring for Environment and Security (GMES) programme in Europe is aimed at using data from space for a wide range of applications for the benefit of society, and this has led to the Sentinel programme, discussed further below. National space programmes may be motivated by national pride as well as national requirements; Nigeria sees a space programme as a means of promoting a science programme in the country, but also needs imagery for development programmes and national requirements such as land distribution.

In the commercial arena the Disaster Monitoring Constellation (DMC) uses satellites built and operated by SSTL. Data can be collected from all of these satellites in the event of a disaster, but the data can also be sold by SSTL by arrangement with the owners.

8.2 Future developments

8.2.1 Future missions and sensors

There are a number of satellites in various stages of planning; these are shown in Table 8.1. GeoEye-2 will be similar to GeoEye-1 but with higher resolution, although still only selling images at 0.5 m GSD. It will have the facility to acquire as many as 40 targets with a size of 2 × 2 nautical miles, as well as large areas, in a single pass. Pleiades will have a GSD of 0.5 m panchromatic and 2 m multispectral, and a swath of 20 km; a 0.5 m GSD will be reached by staggered CCD lines. One Pleiades satellite will have an imaging capacity of approximately 1 000 000 km^2 per day. The satellites will be very agile, having a roll and pitch of 5° in 8 seconds, 10° in 10 seconds and 60° in 25 seconds, which lies between the agility of GeoEye-1 and WorldView. Pleiades can image an area of 100 km × 200 km in a single pass, viewing about 10 targets with its sensors pointed less than 20° off track. Stereo pairs, and also an additional vertical image between fore

Table 8.1 A selection of future satellite missions

Sensor or satellite	Agency country launch date	Platform altitude (km)	GSD at nadir pan/Ms (m)	Swath width at nadir (km)	Channels wavelength (μm)
Pleiades 1 and 2	ASTRIUM Europe 2011/2014	694	0.5/2	20	pan 0.48–0.83 MS: 0.43–0.55, 0.49–0.61,0.60–0.72, 0.75–0.95
NigeriaSat-2	NASRDA, Nigeria, 2011	700	2.5/5	20	pan 0.520–0.900 MS: 0.455–0.520, 0.525–0.600, 0.60–0.72, 0.75–0.95
GeoEye-2	GeoEye, USA 2013		0.25		
SPOT-6 SPOT-7	ASTRIUM Europe	694	1.5/6	60	pan 0.45–0.745 MS: 0.45–0.52, 0.53–0.59,0.635–0.690, 0.775–0.900
Cartosat-3	ISRO India 2013	450	0.33	15	pan 0.5–0.75
KOMPSAT-3	KARI Korea 2011	685	0.7/2.8	16.8	pan 0.45–0.9 MS: 0.45–0.52, 0.52–0.60, 0.63–0.69, 0.76–0.90
ALOS-3	JAXA, Japan 2015	618	0.8/5	50	4 MS bands 185 HS
CBERS-3/4	CRESDA/INPE, China/Brazil 2012/2014	778	5/10	60	pan 0.51–0.85 MS: 0.52–0.59, 0.63–0.69, 0.77–0.89

and aft images, will be taken from the same track. Five panchromatic TDI arrays, each with 6000 pixels, will be combined in one line. Image location accuracy will be 3 m without ground control. India will continue its programme with Cartosat-3, having a planned 0.33 m GSD, as will South Korea with KOMPSAT-3 and JAXA with ALOS-3. The CBERS programme will continue with CBERS-3 and CBERS-4, again with an improved GSD.

8.2.2 Small satellites

The growing market in small satellites is likely to provide more satellites in the future. SSTL, in the UK, offers a range of satellites as shown in Table 8.2, and is primarily concerned with building and operating satellites, not with selling data or products, although images are sold through DMCii, which operates satellites owned by China, Nigeria, Algeria and the UK. The company is able to keep the price down by producing systems designed through experience and using off-the-shelf components, without offering, for example, high positioning accuracy.

Small satellites are also built by other groups. The RapidEye configuration as well as the EROS-satellites belong to the category of small satellites.

8.2.3 Coordinated missions

There are other developments which at present do not involve high resolution optical sensors, but this may change in the future. The European Space Agency (ESA) has initiated the Sentinel programme of missions to support the GMES programme. Sentinel-2 is a polar-orbiting, multispectral high resolution imaging mission for land monitoring and will also deliver information for emergency services. In essence, Sentinel-2 combines a large swath, frequent revisit, and – using a large number of spectral bands – systematic acquisition of all land surfaces at high spatial resolution, all of which makes this GMES mission unique. The first Sentinel-2 satellite is planned for launch in 2013 and will carry an optical payload with visible, near-infrared and

Table 8.2 SSTL products

	Weight (kg)	GSD pan/MS (m)	Comment
SSTL 150		2.5/5	Agile
SSTL 300	300	1/2	Fast slew agility
SSTL 300 S1	350	0.75/1.5	Agile, planned for use in DMC-3 constellation
SSTL ART	500	0.6/2.4	Designed for global coverage at high resolution and low cost

short-wave infrared sensors comprising thirteen spectral bands: four bands at 10 m, six bands at 20 m and three bands at 60 m. The images will be available free of charge.

As part of the Global Earth Observation System of Systems (GEOSS), the Committee on Earth Observation Satellites (CEOS) has established virtual constellations in order to have coordinated Earth observation missions to cover defined requirements. Six constellations currently exist: Land Surface Imaging, Ocean Surface Topography, Atmospheric Composition, Precipitation, Ocean Colour Radiometry and Ocean Surface Vector Winds. CBERS missions are included in the Land Surface Imaging constellation.

The DMC currently has access to satellites owned by national organisations, but SSTL is planning a new constellation which will be operated by a single organisation and will include optical sensors (see Table 8.2) as well as radar sensors. One possible scenario is for wide-swath optical sensors with a morning overpass to image possible targets, to be followed up in the afternoon by a radar sensor with a possible 3 m resolution, providing high resolution imagery independent of cloud.

8.2.4 Applications

It is difficult to predict new applications for high resolution satellite data beyond those discussed in Section 8.1.3, although it can be assumed that these applications will continue to develop. Areas such as marine surveillance may grow, and agricultural monitoring may be used more and more in the future.

8.3 Conclusions

It is clear that high resolution optical sensor technology has reached a stage of maturity and that current trends are towards lower costs, faster access to any point on the Earth and more frequent global coverage. Improved resolution and greater accuracy in positioning are desirable, but expensive. There appears to be a growing public appetite for imagery: satellite systems can provide that quickly, and communication systems can deliver the data to an ever-widening community of users.

Index